Blowout and Well Control Handbook

Blowout and Well Control Handbook

Editor

Mithilesh Kumar

scitus
academics

Blowout and Well Control Handbook

Edited by **Mithilesh Kumar**

Printed in 2017

ISBN: 978-1-68117-338-2

Library of Congress Control Number: 2015939250

© 2016 by

SCITUS Academics LLC,
616, Corporate Way, Suite 2, 4766,
Valley Cottage, NY 10989

www.scitusacademics.com

Contents

vi

Preface

Well control problems are always interesting. The raw power that is released by nature in the form of an oil or gas well blowing out of control is awesome. Well control is one thing and WILD well control is something else. There will be well control problems and wild wells as long as there are drilling operations anywhere in the world. There are some among us that think these problems are always the consequence of some error and can be eliminated. I don't think so. I've seen some that I don't think anyone could have avoided. These problems are part of the business and just go with the territory. The consequences of failure are severe. Even the most simple blowout situation can result in the loss of millions of dollars in equipment and valuable natural resources. These situations can also result in the loss of something much more valuable—human life.

Editor

Chapter 1

Research on Performance of H$_2$ Rich Blowout Limit in Bluff-Body Burner

Hongtao Zheng[1], Yajun Li[1], and Lin Cai[2]

[1]College of Power and Energy Engineering, Harbin Engineering University, Harbin 150001, China

[2]Department of Power and Energy, China Ship Development and Design Center, Wuhan 430064, China

ABSTRACT

In order to investigate H$_2$ rich blowout limit at different blockage ratios and flow velocities, a CFD software FLUENT was used to simulate H$_2$ burning flow field in bluff-body burner, and the software CHEMKIN was adopted to analyze the sensitivity of each elementary reaction. Composition Probability Density Function (C-PDF) model was adopted to simulate H_2 combustion field in turbulence flame. The numerical results show that reactions R2 and R9 possess the largest positive and negative temperature sensitivity. Temperature has a very important influence on these two reactions. When equivalence ratio is 1, the mixture is most ignitable, and the critical ignition temperature is 1550K. There should be an optimal blockage ratio which can stabilize the flame best. When the blockage ratio remains unchanged, the relationship between H_2 RBL and flow velocity is a logarithmic function. When the flow velocity remains unchanged, the relationship between H_2 RBL and blockage ratio is a parabolic function. A complete extinction requires three phases: the temperature sudden decline in the main stream, the energy dissipation from the recirculation zone to the main stream, and the complete extinction of the flame.

INTRODUCTION

Bluff-body stabilized combustion with triangular or cone stabilizers is common in afterburners of military aircraft. A central recirculation zone (CRZ) will form in the wake of the bluff-body burner [1]. The heat will diffuse to the main stream from the flame frontier. The entrainment of hot gases will improve the combustion stabilization. If the fuel concentration is ultralean or ultrarich, the heat released from the flame frontier cannot compensate that of dissipation to the main stream, and then the temperature will decrease gradually, finally inducing extinction.

Lots of researches on flame stabilized mechanism in a bluff-body burner have been carried out both in terms of experiment and theoretical treatment. Experimental researches on this problem are extremely important, but a large-scale systematic mechanism analysis via experiments is both expensive and time consuming. The Volvo Aero Corp. [2] carried out a lot of experiments on triangular bluff-body stabilized combustion rig. Shanbhogue et al. [1] found that the flame instability is dominated by the lower intensity and the convective instability of the shear layer. He put forward that blow off will occur in multiple steps: local extinction along the flame sheer, large-scale wake disruption, and a final blow off. Frolov et al. [3] formulated a flame stabilization criterion called Michelson Criterion, according to this criterion, a flame will be blown off from the flameholder when Michelson number is <1; his result shows that there will be an optimal flame-holder size at which the best stabilization parameters were achieved. Wright [4] performed lots of experiments to define the influence of blockage on flame stabilization by bluff-bodies in ducted flow. His experiments indicated that the length of the recirculation zone varies inversely as the square root of the blockage and the flow speed past the wake increases almost linearly with blockage. He found that while the combustion was taking place, the flow speeds and flame geometry depended on the blockage ratio. However, at the flame blow off, the characteristic mechanical time is independent of that. The most important conclusion gained by Wright is that the maximum blow off speed occurs at a relatively low blockage ratio. Dawson et al. [5] found that blow off is approached by increasing the bulk velocity or decreasing the equivalence ratio. Griebel et al. [6] and Schefer [7] found that the maximum blowout velocity occurred at stoichiometric conditions. Barlow et al. [8] made use of an experimental method to study the importance of molecule diffusion and turbulence transport on flame structure. His study showed that there will be an evolution in those flames from a scalar structure dominated by molecular diffusion to one dominated by turbulent transport with Re increasing.

On the other hand, computational fluid dynamics (CFD) has been widely used to study the turbulent reacting flows, fluid machinery, and combustion systems to predict device performance and optimize their structures. Many experiment studies are used to validate the simulation accuracy and to explain the flame extinction mechanism. For example, Giacomazzi et al. [9] tested the applicability of a sub grid scale Fractal Model for LES (FM-LES) simulation of turbulent combustion by simulating a bluff-body premixed flame anchored in a straight channel. Eugenio found that 3D vortex structures periodically shortening the recirculation zone downstream of the bluff-body and entraining fresh mixture into the hot zone, this physical mechanism is involved in flame anchoring. Eriksson [2] investigated Zimont Turbulent Flame Closure Model (TFC) in conjunction with different turbulent models in simulating premixed bluff-body stabilized flame. And he found that the TFC model combined with k-ω model accurately captures the recirculation zone length and overall turbulent flame speed, the combined effect is not captured well in steady state RANS. Lin and Holder [10] studied the effects of inlet turbulent intensity and angle of attack on the chemically reacting turbulent flow and thermal fields in a channel with an inclined bluff-body flame holder. Sjunnesson [2] reported the computation of the triangular bluff-body stabilized combustion using a two-step reaction solved with Arrhenius Expression in conjunction with the Magnusen-Hjerthager combustion model and k-epsilon turbulent model. Kim et al. [11] found that LES modeling approach can reproduce the variation of recirculation zone length while the equivalence ratio changed. This approach was successfully used to assess the lean blowout condition and evaluate its behavior and physics of combustion instability. Jones and Prasad [12] adopted C-PDF/LES model to exhibit the local extinction and re-ignition in turbulent combusting and to describe the interaction between turbulence and combustion. His numerical result was in good agreement with American Sandia Flame experiment data.

Even though there is a recirculation region in a bluff-body burner, the extinction will still occur if the stabilized ignition point was blown to the outside of CRZ. CRZ takes a very important effect on flammability. So the aim of the present work is to study the influence of flow velocity and blockage ratio on H_2 Rich Blowout Limit (RBL) and finally summarize a formula for H_2 RBL.

GEOMETRY AND MATHEMATICAL MODEL

Figure 1 shows the geometry of the burner with cone bluff-body and straight channel. To save the calculation expend, 2-dimension axis-symmetry model was used. Figure 2 shows the mesh adopted for the calculation domain, and the total grid number is 6.0e + 04.

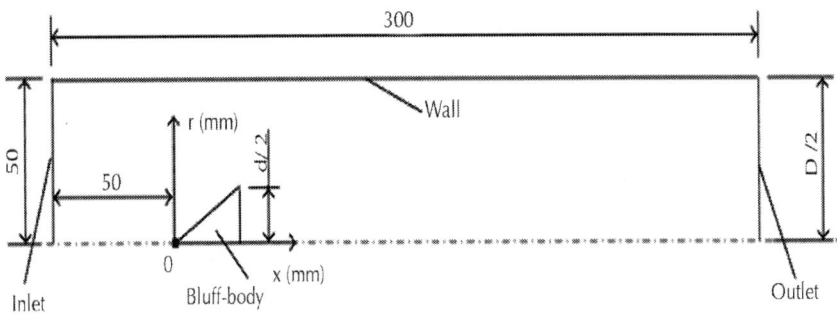

Figure 1: Geometry structure of bluff-body burner.

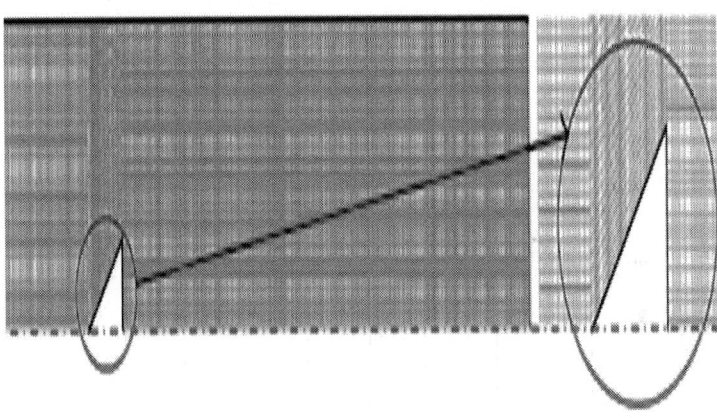

Figure 2: The calculation mesh.

Boundary conditions: mixture inlet temperature is 293K, inlet pressure is 1atm and inlet velocities are shown in Table 1. The mixture is made up of H$_2$ and air, and the concentration of H$_2$ is shown in Table 5.

- Outlet: pressure outlet
- Wall: adiabatic boundary

Table 1: Blockage ratio and gas flow velocities (1atm, 293K)

Bluff-body diameter d/mm	20	30	40	50	60	70
Blockage ratio B	0.2	0.3	0.4	0.5	0.6	0.7
Gas velocity V/ (m/s)	1	2	5	10	20	50
Re_D	4.0e+3	8.3e+3	2.2e+4	4.4e+4	9.1e+4	2.4e+5

The computations are repeated for different combinations of gas velocity and blockage ratio. The definition of Reynolds number based on the channel width has been given out as follows:

$$Re_D = \frac{\rho u D}{\mu},$$

$$(2.1)$$

where ρ is mixture density, u is mixture velocity, D is Channel width, μ-mixture viscosity.

The blockage ratio B is defined as

$$B = \frac{d}{D}.$$

$$(2.2)$$

In combustion flows, conservation equations for mass, momentum, energy, and species are solved. The standard k-ε and LES models were adopted, respectively, to simulate the turbulence flow in conjunction with C-PDF and Eddy-Dissipation-Concept (EDC) combustion mode.

Composition PDF Transport Equation (C-PDF) is as follows:

$$\frac{\partial}{\partial t}(\rho P) + \frac{\partial}{\partial x_i}(\rho u_i P) + \frac{\partial}{\partial \psi_k}(\rho \omega_k P) = -\frac{\partial}{\partial x_i}\left[\rho \left\langle u_i'' \mid \psi \right\rangle P\right] + \frac{\partial}{\partial \psi_k}\left[\rho \left\langle \frac{1}{\rho}\frac{\partial J_{i,k}}{\partial x_i} \mid \psi \right\rangle P\right].$$

$$(2.3)$$

The two terms on the right-hand side represent the PDF change due to scalar convection by turbulent scalar fluxal and molecular mixing/diffusion, respectively.

The flow field researched in this paper is turbulence refer is to Re_D as shown in Table 1. Because the reaction rate highly nonlinear, modeling the mean reaction rate in a turbulent flow is extremely difficult. C-PDF is an alternative effective method to solve the premixed combustion in a turbulent flow. The principal strength of C-PDF transport approach is that the highly nonlinear reaction term is completely closed and requires no modeling. The turbulent scalar flux term is modeled by gradient-diffusion assumption. The molecular mixing/diffusion is modeled by MC, IEM, or EMST models [13]. C-PDF transport model adopts a detailed chemical mechanism for modeling the reaction rate in a turbulent flame. With an appropriate chemical mechanism, kinetically controlled species such as CO and NOx, as well as flame ignition and extinction, can be predicted by C-PDF model. Bisetti and Chen [14] adopted Join-PDF/LES approach to research Sandia Flame D, and their result showed that the prediction by EMST is quite accurate near stoichiometric, but overall trend remains unpredicted at other conditions. While Lindstedt et al. [15] found that the numerical result is in good agreement with the experiment data by MC molecular mixing model, if it turns to IEM model, it cannot capture extinction and reignition. Finally, MC molecular mixing model was used in this paper.

C-PDF transport equation cannot be solved by finite volume method; a Lagrangian Monte Carlo [13] method has been used to solve it. Because time scales of some reactions are very fast, while others are very slow, disparity of time scales results in numerical stiffness problem. It means that extensive computational load is required to integrate the chemical source term. To solve the numerical stiffness problem, in-situ adaptive tabulation (ISAT) has been employed to dynamically accelerate the chemistry calculations. Correa and Pope [16] made use of this method to calculate the burning process for one bluff-body burner, and the numerical result was in coincidence with the experiment data.

Table 2 gives out the detailed chemical mechanism of hydrogen adopted in this paper.

Table 2: Hydrogen chemistry reaction

No.	Reaction	A_i	b_i	E_i
1	$O_2 + H \Rightarrow OH + O$	2.000E+14	0.00	70.30
2	$OH + O \Rightarrow O_2 + H$	1.568E+13	0.00	3.52
3	$H_2 + O \Rightarrow OH + H$	5.060E+04	2.67	26.30

4	OH + H $\Rightarrow$ H$_2$ + O	2.222 E+04	2.67	18.29
5	H$_2$ + OH $\Rightarrow$ H$_2$O + H	1.000 E+08	1.60	13.80
6	H$_2$O + H $\Rightarrow$ H$_2$ + OH	4.312 E+08	1.60	76.46
7	OH + OH $\Rightarrow$ H$_2$O + O	1.500 E+09	1.14	0.42
8	H$_2$O + O $\Rightarrow$ OH + OH	1.473 E+10	1.14	71.09
9	O$_2$ + H + M $\Rightarrow$ HO$_2$ + M	2.300 E+18	−0.80	0.00
10	HO$_2$ + M $\Rightarrow$ O$_2$ + H + M	3.190 E+18	−0.80	95.39
11	HO$_2$ + H $\Rightarrow$ OH + OH	1.500 E+14	0.00	4.20
12	HO$_2$ + H $\Rightarrow$ H$_2$ + O$_2$	2.500 E+13	0.00	2.90
13	HO$_2$ + OH $\Rightarrow$ H$_2$O + O$_2$	6.000 E+13	0.00	0.00
14	HO$_2$ + H $\Rightarrow$ H$_2$O + O	3.000 E+13	0.00	7.20
15	HO$_2$ + O $\Rightarrow$ OH + O$_2$	1.800 E+13	0.00	−1.70
16	H + H + M $\Rightarrow$ H$_2$ + M	1.800 E+18	−1.00	0.00
17	OH + H + M $\Rightarrow$ H$_2$O + M	2.200 E+22	−2.00	0.00
18	O + O + M $\Rightarrow$ O$_2$ + M	2.900 E+17	−1.00	0.00
M	H$_2$O/6.5 O$_2$/0.4 N$_2$/0.4/		Third body efficiency	
Unit	$A_i - cm.mole.s.k, \beta_i - 1, E_i - kJ / mole.$			

VALIDATIONS OF MATHEMATICAL MODEL

Validations of Independence of Grid Size and Time Step

The studies on grid size and time step independence have been performed to determine the optimal grid and time step with a good accuracy for the simulation. Also, the k-epsilon-C-PDF model was used.

Table 3 gives out the grid size range which changes from 0.5mm to 2.0mm (the grid number varies from 1.0e + 05 to 1.6e + 04). The time step is, respectively, 0.05ms, 0.10ms, 0.20ms, and 0.50ms, as shown in Table 4. The average temperatures on section x=150mm and x=350mm were adopted to verify the accuracy of grid size and time step. Table 3 shows that on section x=150mm, the average temperature at grid size Δ=1.0mm is only 1°C higher than that at grid size Δ=0.5mm. On section x = 350 mm, the average temperature at grid size Δ = 1.0 mm is 13°C higher than that at grid size Δ = 0.5 mm. So, the numerical simulation is independent when grid size is Δ = 1.0 mm the grid number is (6.0e + 04). Table 4 shows that on section x = 350 mm, the maximum temperature error is only 20°C between time step 0.1 ms and 0.05 ms. So, when the time step is 0.1 ms, the numerical result does not rely on it.

Table 3: Grid size independence validation with time step 0.1ms

Grid size/mm	$\overline{T_{x=150}}\,/\,k$	DT/K	$\overline{T_{x=350}}\,/\,k$	DT/K	Grid amount
0.5	419	—	768	—	1.0e + 05
0.8	420	1	760	8	9.2e + 04
1.0	420	1	755	13	6.0e + 04
1.5	415	4	707	61	2.8e + 04
2.0	414	5	666	102	1.6e + 04

Table 4: Time step independence validation with grid size 1.0mm

Time step/ms	$\overline{T_{x=150}}\,/\,k$	DT/K	$\overline{T_{x=350}}\,/\,k$	DT/K	Remark
0.05	424	—	775	—	—
0.10	419	5	755	20	Independent
0.20	405	14	707	68	Worse
0.50	402	22	604	171	Worse

Table 5: H$_2$ RBL (volume concentration)

V	B					
	0.2	0.3	0.4	0.5	0.6	0.7
1	0.775	0.782	0.783	0.786	0.779	0.777
2	0.765	0.775	0.774	0.776	0.770	0.767
5	0.753	0.763	0.763	0.765	0.757	0.754
10	0.745	0.755	0.753	0.754	0.749	0.746
20	0.735	0.743	0.743	0.745	0.740	0.737
50	0.723	0.728	0.728	0.731	0.725	0.724

Figure 3 shows the temperature curves at different grid size and time step. It indicates that when the grid size is 1.0 mm and the time step is 0.1 ms, the temperature error between the numerical result and the experiment data is extremely small. So, it can be included from the results that the optimal grid size is Δ = 1.0 mm and the optimal time step is Δt = 0.1 ms.

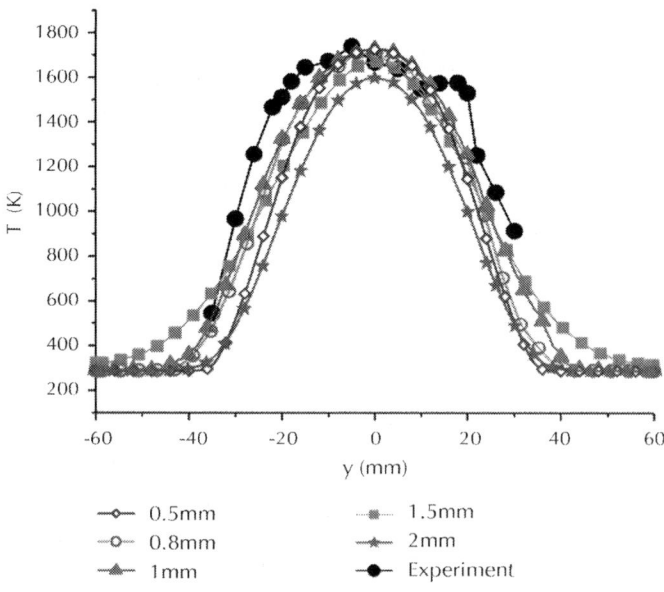

(a)

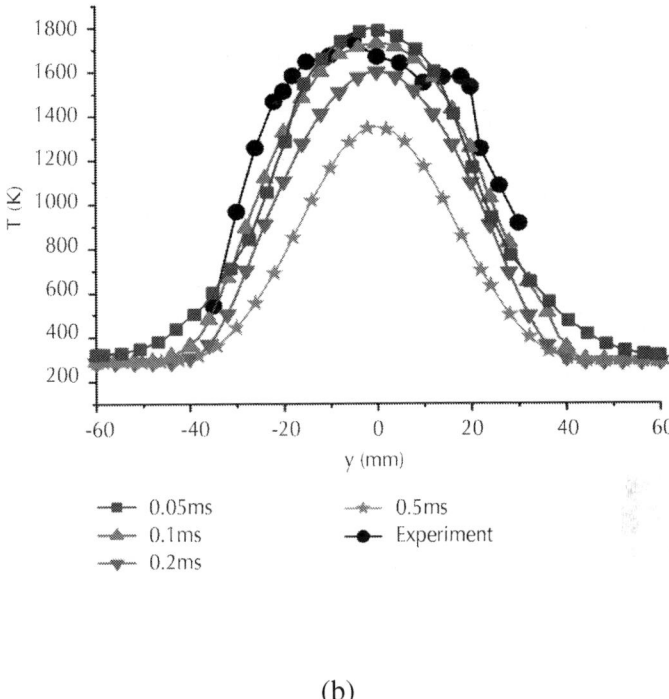

(b)

Figure 3: Profiles of temperature on section mm at different grid size and time step. (a) Grid size independence and (b) Time step independence.

Experiment Validation for Model Accuracy

Figure 4 shows the combustion device of Volvo Aero Corp. triangular bluff-body which has been widely used to research the flame stabilization mechanism both in terms of experiments and theoretical data. Over the years, many CFD researches relied on the experiment data of this device have been carried out to investigate its stabilized mechanism [2]. In order to validate the accuracy of the mathematical model, this combustion rig was used with Smagorinsky-Lily-LES-EDC (SL-LES-EDC) model, k-epsilon-EDC model, and k-epsilon-C-PDF model.

(a)

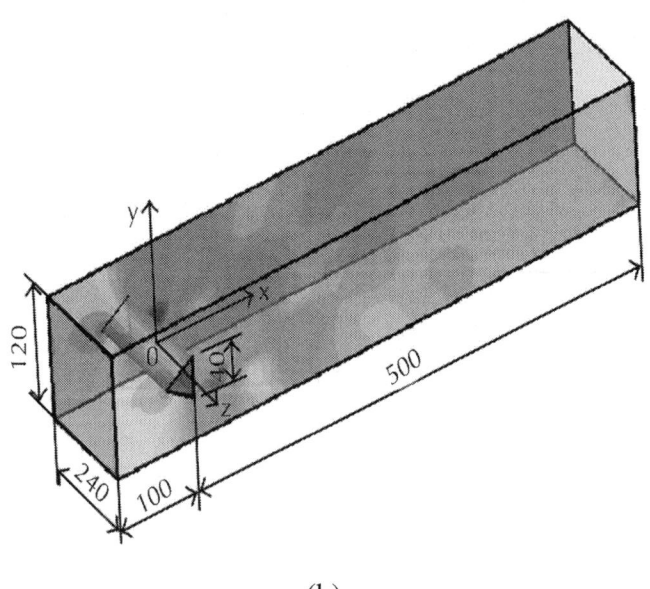

(b)

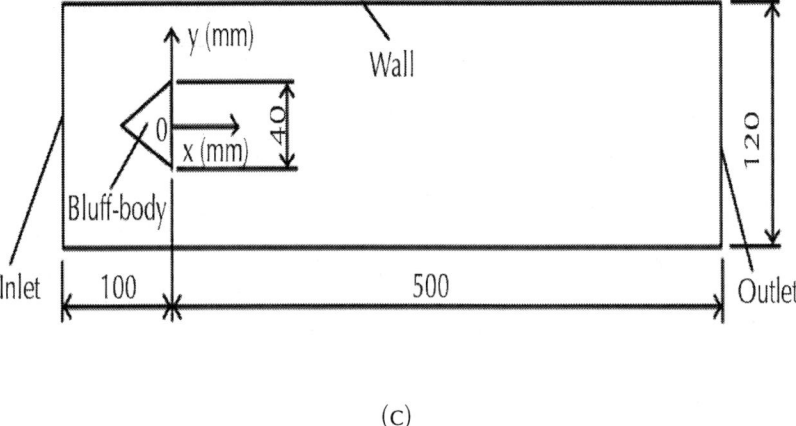

(c)

Figure 4: Validation rig and calculation zone geometry. (a) Validation rig [2], (b) Three-dimensional calculation zone and (c) Vertical section—XOY

Geometry model and boundary conditions are as follows:

? Length × width × height = 660 × 240 × 120 mm;

? Side length of bluff-body with the equilateral triangular cross-section: 40mm;

? Inlet condition of mixture of air and propane: T = 288 K, V = 17 m/s, p = 1 atm,

? mass flow rate is 0.6 kg/s, equivalence ratio is φ = 0.65;

? Outlet: pressure outlet;

? Wall: adiabatic boundary;

? Fuel oxidation was modeled by one-step global reaction:

$$C_3H_8 + 5O_2 \Rightarrow 3CO_2 + 4H_2O$$

$$(3.1)$$

The reaction rate is that proposed by Fluent Database according to Arrhenius Law [9].

$$\dot{\omega} = \frac{d[C_3H_8]}{dt} = -4.836 \times 10^{-9} e^{-15100/T} \times [C_3H_8]^{0.1}[O_2]^{1.65}.$$

$$(3.2)$$

The Reynolds number based on the bluff-body burner device and on the velocity at the bluff-body location is about 10^5. The flow field is simulated using compressible N-S equations.

Figures 5, 6, and 7 show the comparison between numerical results and the experiment data. The results indicate that SL-LES-EDC model will overpredict the recirculation length and underestimate its width. Giacomazzi et al. [9] found that FM-LES-EDC model can accurately predict the recirculation zone position, but in combustion flow field, it will underestimate the velocity nearby the channel wall, and it may be because of the disadvantage of the FM-LES-EDC in dealing with the turbulent viscosity nearby the channel wall. Relatively, the k-epsilon-C-PDF, k-epsilon-EDC, and SL-LES-EDC models can predict the velocity better than FM-LES nearby the channel wall.

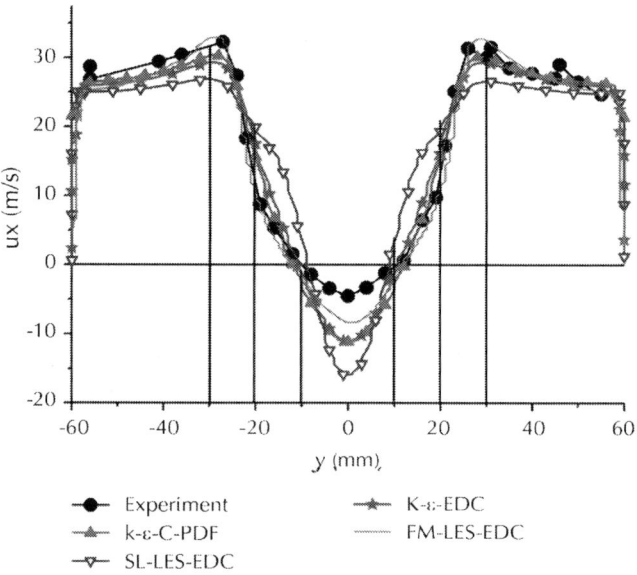

(a)

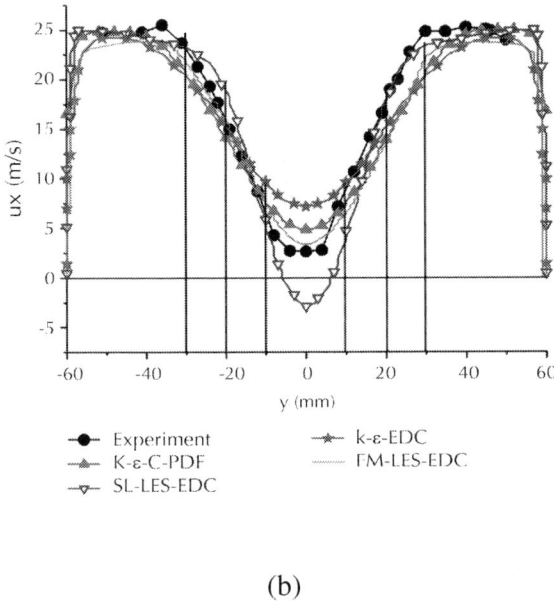

(b)

Figure 5: Profile of U$_x$ of cold field at sections (a) Section x=15 mm, (b) Section x=61mm

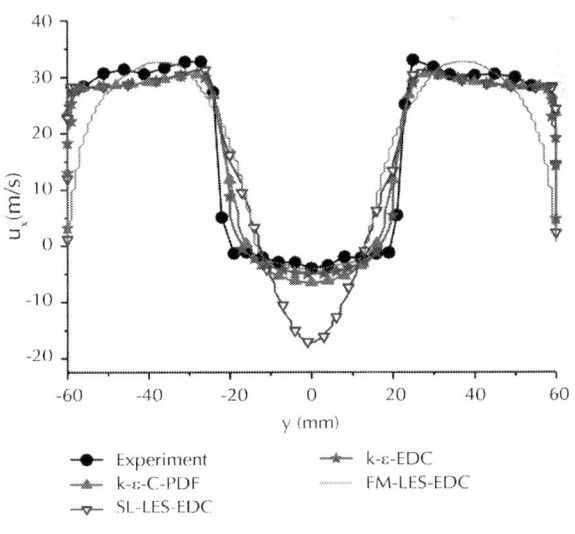

(a)

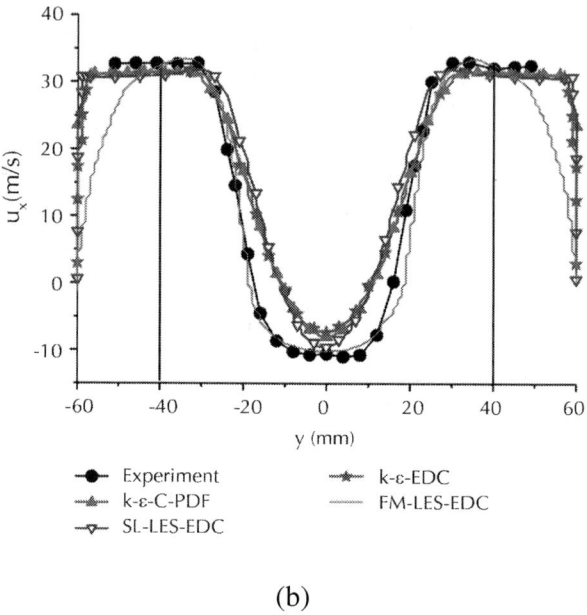

(b)

Figure 6: Profile of U_x of combustion field at sections (a) x=15 mm, (b) x= 61mm.

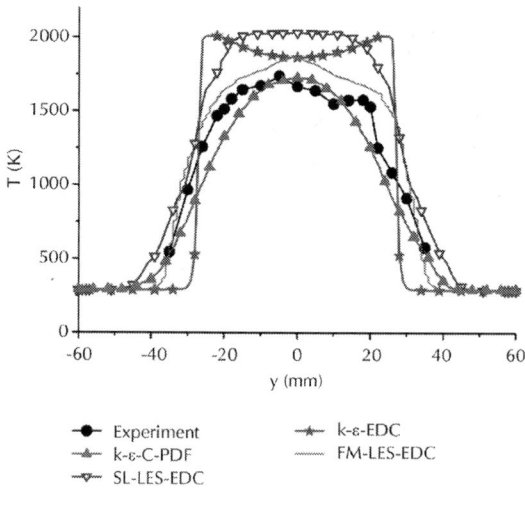

(a)

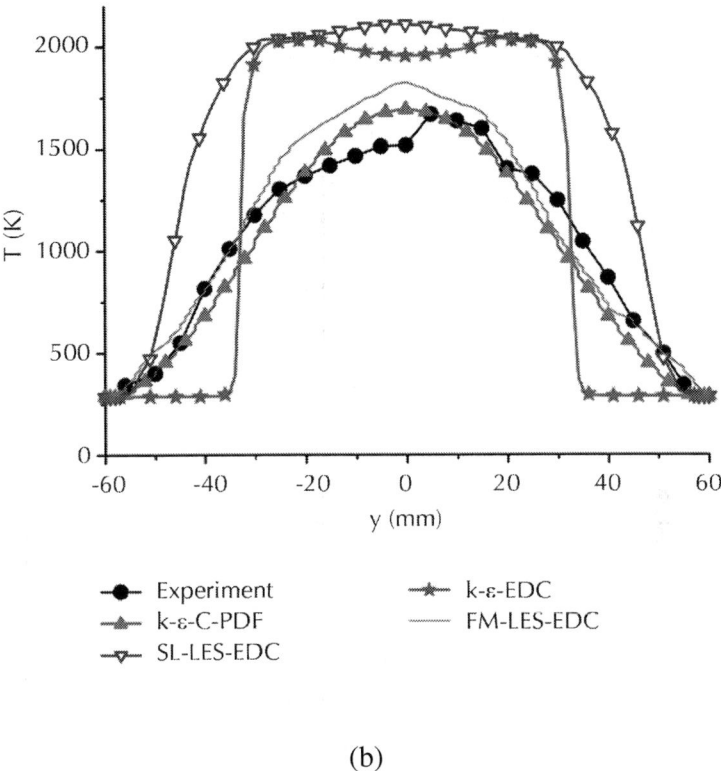

(b)

Figure 7: Profiles of temperature at sections (a) x= 150 mm, (b) x= 350 mm.

Figures 7 and 8(b) show that in reacting case, SL-LES-EDC model will overestimate the flame width and the temperature in the flame zone. The maximal temperature of the flame from SL-LES-EDC is 400K higher than that of the experiment data. The k-epsilon-EDC model will also overestimate the temperature in the flame center, but it will underestimate the flame temperature at the outside of the flame zone (see Figures 7and 8(c)). Figure 7 shows that FM-LES-EDC can relatively accurately predict the flame temperature, except for the maximal temperature, and it will overestimate its value about 100K. In spite of this situation, it can reproduce the flame vortex structures and the flame periodic fluctuation. Compared with FM-LES-EDC model, k-epsilon-C-PDF can predict the flame temperature more accurately but it cannot capture the flame vortex shell accurately.

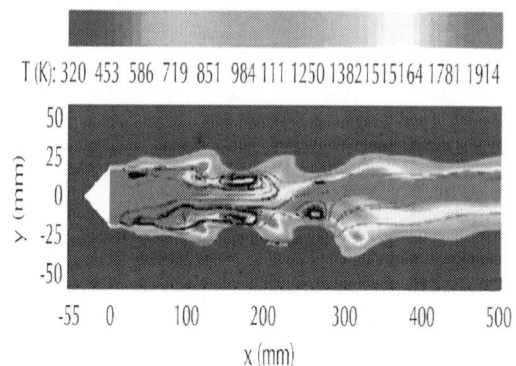

(a)

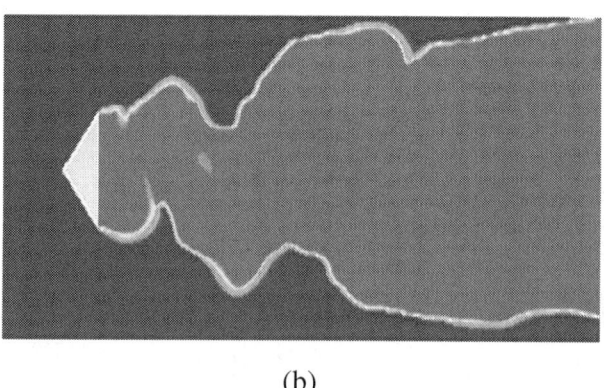

(b)

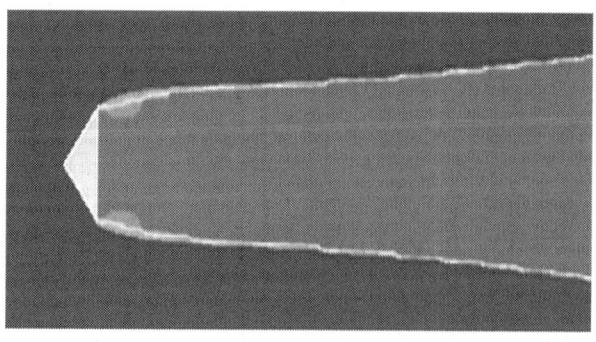

(c)

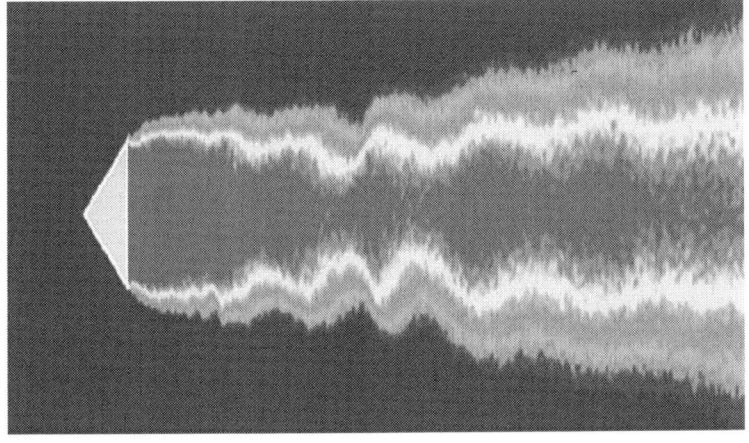

(d)

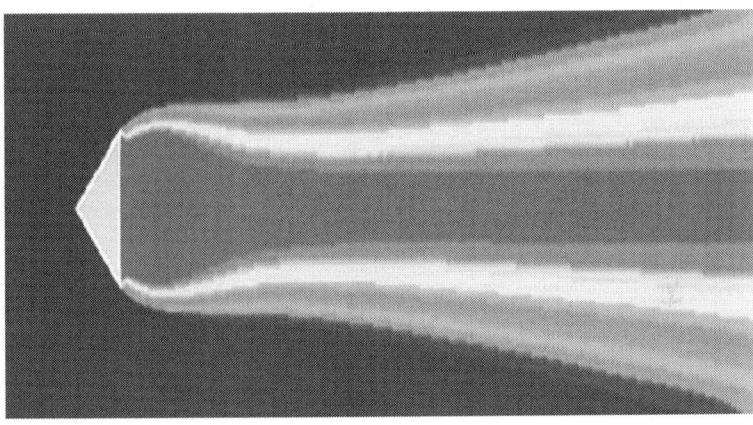

(e)

Figure 8: Comparison of temperature field by different mathematical model. (a) Temperature field from Eugenio by FM-LES-EDC [9], (b) Instantaneous temperature by SL-LES-EDC model, (c) Instantaneous temperature by k-epsilon-EDC model, (d) Instantaneous temperature by k-epsilon-C-PDF model, and (e) Mean temperature by k-epsilon-C-PDF model of 10 periodic.

In a word, the agreement observed between k-epsilon-C-PDF model result and published classics experiment data is acceptable. The k-epsilon-C-PDF combustion model can accurately predict the flame temperature, while SL-LES-EDC model can accurately predict the vortex structures and explain the extinction mechanism.

Figure 9 shows the comparison of recirculation zone between cold field and combustion field based on k-epsilon-C-PDF model. The figures indicate that the performance of recirculation zone of combustion field is apparently different from that of the cold flow. The recirculation length of combustion field is about two times more than that of the cold flow. This should be attributed to the flame fluctuation and the dramatic heat releasing from chemistry reaction. The flame propagation extends the axis momentum of the velocity and lengthens the recirculation zone further downstream from the bluff-body.

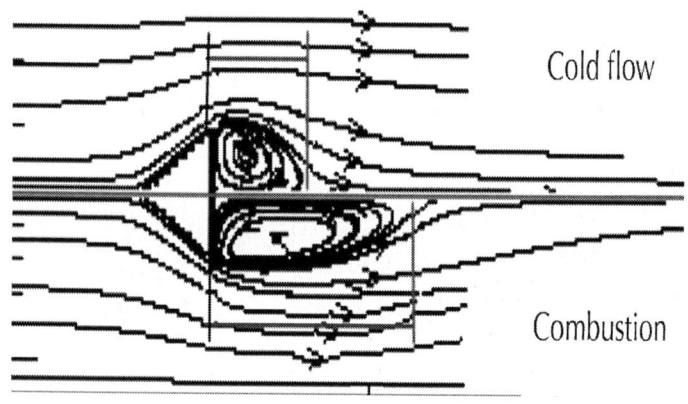

Cold flow

Combustion

(a)

(b)

Figure 9: Comparison of recirculation zone between cold flow and combustion field. (a) Streamline and (b) U$_x$=0 interface

Figure 10 shows the streamlines of cold field and combustion field from SL-LES-EDC model and k-epsilon-CPDF model. The results show that, in the cold field, SL-LES-EDC model can capture the asymmetric von Karman vortex street at the wake zone of the bluff-body. The period of the vortex shedding is 110Hz, in agreement with the measurements (105Hz) [9]. The asymmetric von Karman shedding of coherent vortices no longer exists in the reacting case. Shanbhogue et al. [1] attributed this absence of Karman vortex street for reacting case to the dilatation effect of the heat release. However, what can be concluded from the numerical simulation is that the asymmetric Karman Vortex Street does not disappear in the reacting case, it just converges with the downstream nearby the bluff-body, and the direction of the vortex at the wake zone is still also changed periodically.

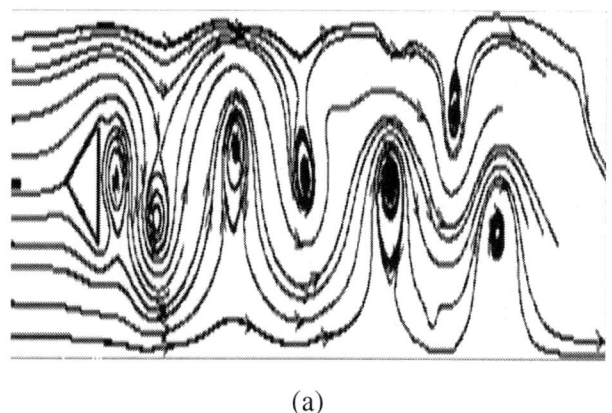

(a)

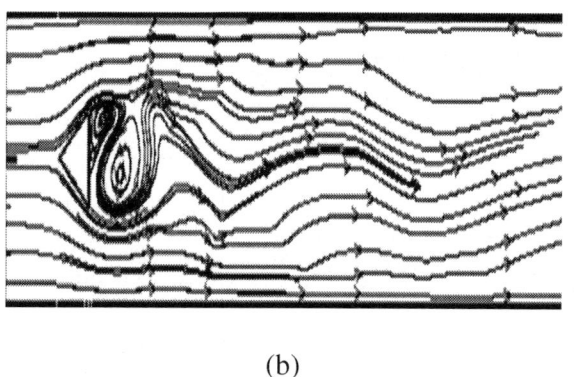

(b)

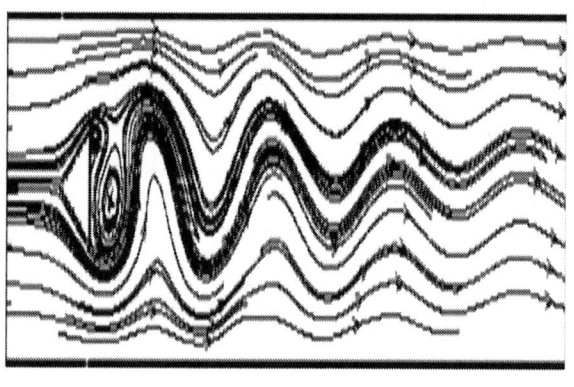

(c)

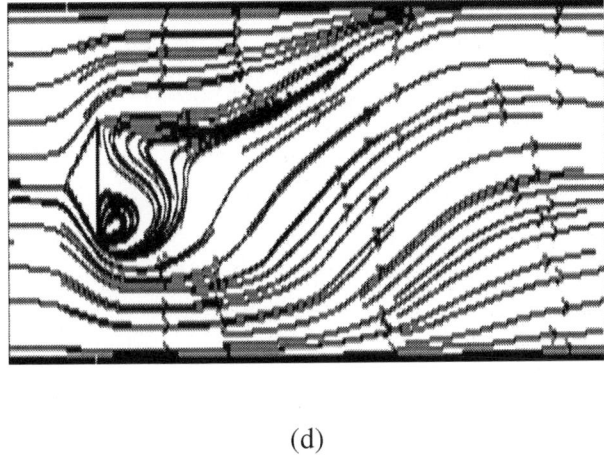

(d)

Figure 10: Streamline of cold flow field and combustion field. (a) SL-LES-EDC cold field, (b) SL-LES-EDC combustion field, (c) k-epsilon-C-PDF cold field and (d) k-epsilon-C-PDF combustion field.

Figure 10 shows that k-epsilon turbulent model cannot capture the asymmetric von Karman vortex street shedding, while the model can just capture the central recirculation zone after the flameholder. Also, in the cold field, k-epsilon model can predict the velocity fluctuation like "a polliwog tail" alternation periodically. That is because the dilatation effect of the heat releases possessing the dominant influence than the fluctuation of Karman Vortex Street. Figure 11 gives out an image of a combustion experiment based on a triangular bluff-body [1].

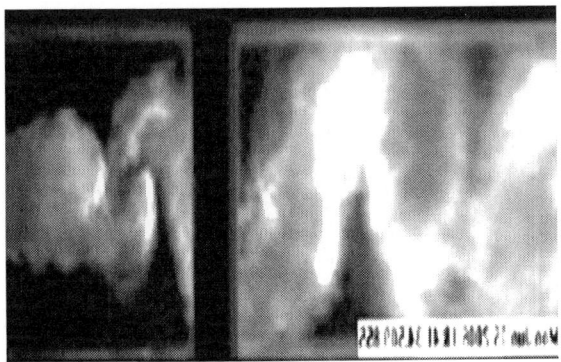

Figure 11: Image of combustion experiment [1].

Figure 12 shows the extinction process of the triangular bluff-body combustion field. What can be included from the numerical is that the entire extinction process experiences three phases: (1) the appearance of discontinuous flame; (2) the flame local extinction and reignition in the recirculation zone; (3) global extinction.

- Discontinuous flame: when the equivalence ration is close to blow off, the flame temperature would decline rapidly, chemistry reaction would be slower, and the heat transfer and dissipation to the flame sheet can ignite the fresh mixture and, finally, induce the discontinuous flame. The first discontinuous position presents at the recirculation zone stagnant point.

- Local extinction and reignition: if the cold mixture is heated up and reignited by flame kernel exactly, the flame is at critical blowout limit state of being acute and unstable. The local extinction and reignition will alternately appear in the recirculation zone. During the blow off, significant fragmentation of the flame occurred, with branches of flame remaining anchored in the bluff-body wake zone.

- Global extinction: the fame pockets moved to the downstream of the recirculation and finally induced the global extinction. Overall blow off occurred with the gradual elimination of these flame fragments and local extinction [1].

Flame fragments

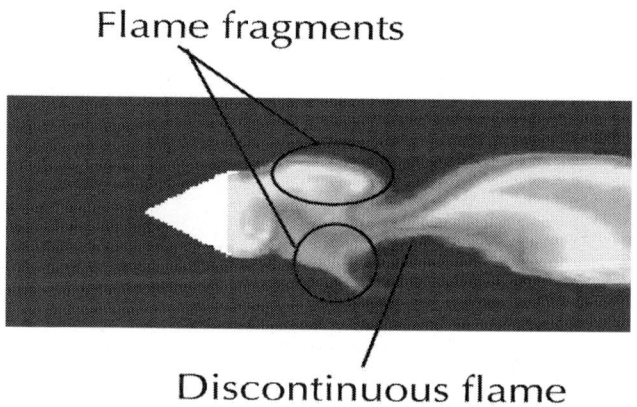

Discontinuous flame

(a)

Flame fragments

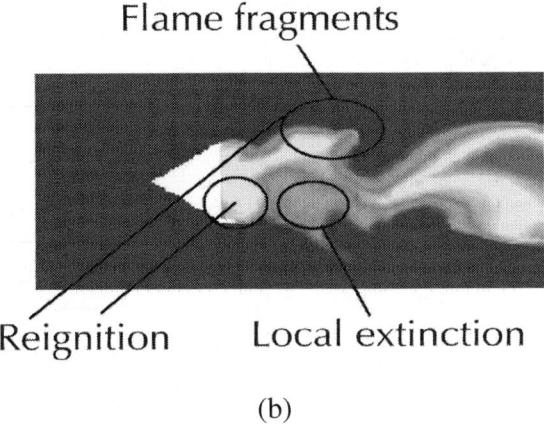

Reignition   Local extinction

(b)

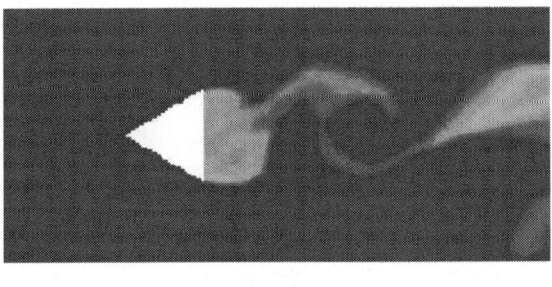

Global extinction

(c)

Figure 12: Extinction process. (a) Discontinuous flame, (b) Local extinction and reignition, (c) Global extinction.

RESULT AND ANALYSIS

Hydrogen Rich Blowout Limit

H_2 RBL is defined as the mole fraction of H_2 in the mixture. It is the H_2 mole concentration limit beyond which the extinction will occur. So, it is dimensionless (%) consider

$$RBL = \frac{n_{H_2}}{n_{mixture}},$$

(4.1)

where, n_{H_2} -mole flow rate of H_2 in mixture, $n_{mixture}$-mole flow rate of mixture. Because the higher H_2 RBL means the wider operation range and the mixture species can change in a wider range. And it is beneficial for the bluff-body burner, so it is interesting to find a way to improve H_2 RBL

Table 5 and Figure 13 show the numerical results of H_2 RBL at different blockage ratio and gas velocities. It can be found that when blockage ratio remains unchanged, H_2 RBL decreases as gas flow velocity is increased. The increasing of the gas velocity will enhance the flow fluctuation and the turbulence intensity, at the same time, the convection and conduction between reactant and product will be improved, and the burning will be enhanced. But when the gas velocity is too large, the flame is close to blowout limit, the turbulence fluctuation will affect the flame's propagation greatly near blowout flames. There will be more and more burning mass which will penetrate into the cooling mixture, at the same time, fresh reactants will penetrate through the CRZ from the flame forward region and abundant of flame fragments will occur, then the separate flame fragments will move randomly inside the CRZ or cooling mixture. When the heat released from chemical reaction is not sufficient to maintain its burning, the flame local extinction will occur. In a word, the main reason of flame blow-out is the generation and elimination of the flame fragments caused by gas velocity increase. So the flame blow out will be approached by increasing the gas velocity, and increasing the mainstream velocity has an adverse effect on flame flammability. This conclusion is accordance with that of EL-feky and Penninger [17] and Dawson's [5] experiment result. Shanbhogue [1] also found that temporally localized extinction occurred sporadically on near blow off flames. Under certain conditions the flame cannot persist indefinitely when Re is too large. The number of local extinction per unit time increase as blow off is approached, and the ultimate blow off event results from more and more local extinction. His conclusion does also support the conclusion summarized in this paper, and the numerical results from the present work are in accordance with his conclusion.

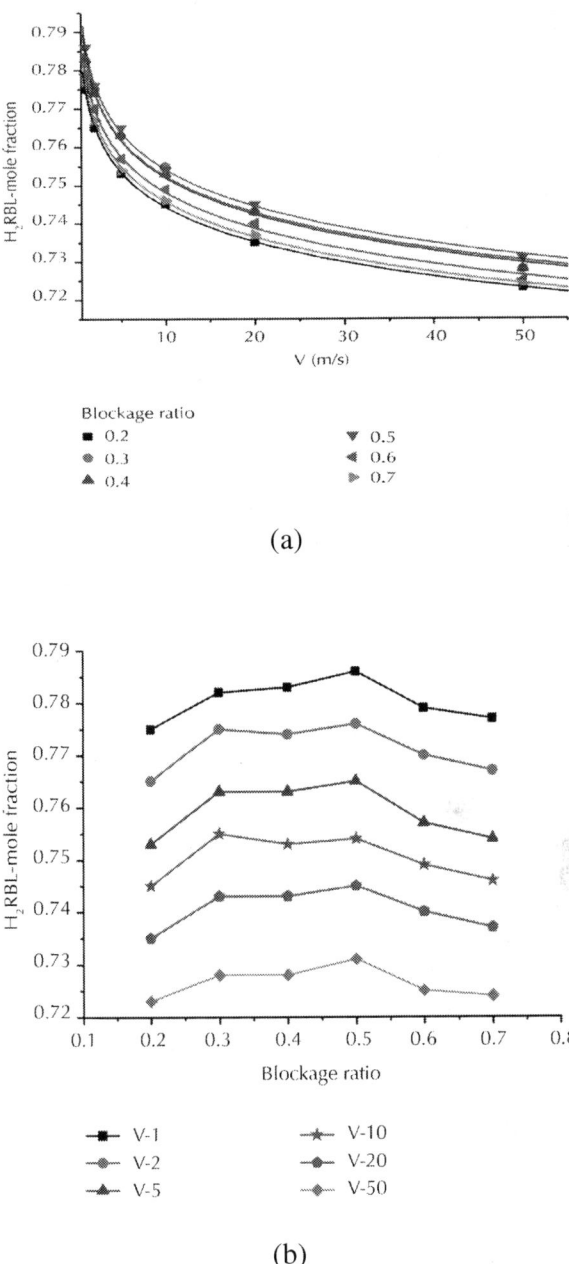

Figure 13: H$_2$ RBL versus gas flow velocity and blockage ratio. (a) Linear co-ordinate and (b) Curves of H$_2$ RBL versus blockage ratio.

Table 5 and Figure 13 also show that when the gas velocity remains unchanged, the H_2 RBL will rise first and decline later as the blockage ratio is increased. It means that there will be an optimal blockage ratio (in the present work, the optimal blockage ratio is B = 0.5, as shown in Figure 13(b)) which can stabilize the flame best. Also at the optimal blockage ratio, the H_2 RBL will reach the largest value. Blockage ratio has a very important effect on the flame geometry, flow speeds, and CRZ length which will influence the flame stability greatly. When the blockage ratio is increased, both the length of CRZ and the gas velocity at the bluff- body channel will increase rapidly. However, they have an opposite effect on H_2 RBL; for example, the H_2 RBL will rise as the length of CRZ is increased, while it will decrease as the gas velocity is increasing. As a result, there would be an optimal blockage ratio. So, the curves in Figure 13(b) are concave. Frolov et al. [3] have carried out the same conclusion about the optimal bluff body diameter by numerical simulation and experiment. Wright [4] found that the maximum blow off speed will occur at B = 0.35 for flat-plate flame-holders. He found that the blockage ratio has a very important effect on the flame geometry, flow speeds, and CRZ length which will influence the flame stability greatly.

In a word, the H_2 RBL will be improved by decreasing the gas velocity, or by increasing the blockage ratio before the optimal value. But if the blockage ratio increases when it has exceeded the optimal value, the H_2RBL will decline. So, when the surroundings remain unchanged in bluff-body burner, the H_2 RBL is a function of gas flow velocity and blockage ratio. To investigate the relationship between gas flow velocity, blockage ratio, and H_2 RBL, the gas velocity has been denoted by logarithmic coordinates (named, logarithmic velocity-lg) as shown in Figure 14(a). It can be seen that the function relationship between H_2RBL and lg is nearly linear. The H_2 RBL will linearly decline with the increase of lg, so the H_2 RBL fitting formula could be assumed as follows:

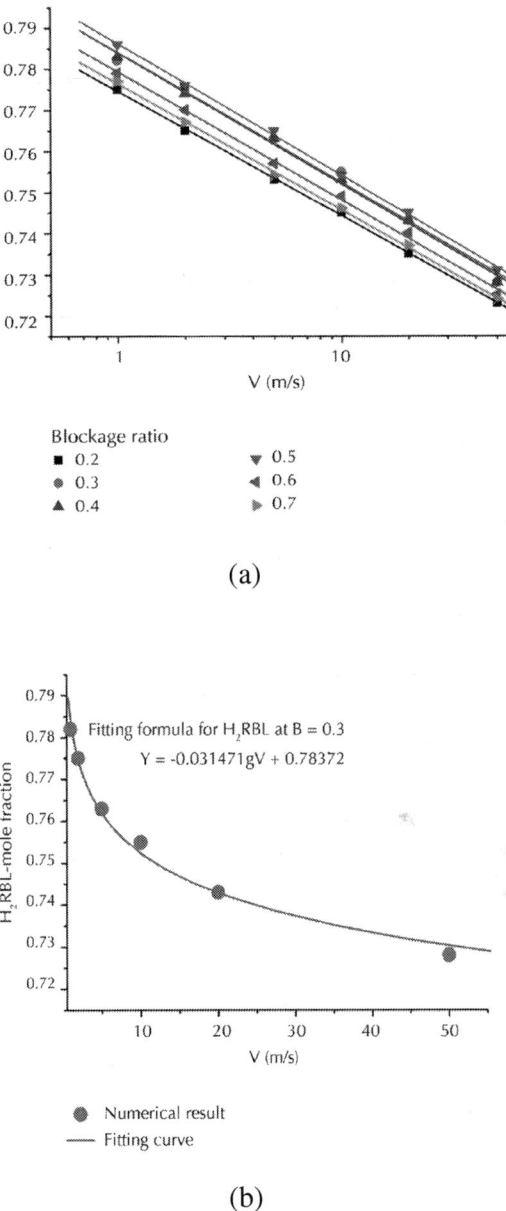

(a)

(b)

Figure 14: Curves of H$_2$ RBL versus gas flow velocity and fitting curve (the symbols are numerical results and the lines originate from the fitting formula). (a) Logarithmic coordinate, (b) Linear coordinate.

Linear fitting of the numerical data results by Least Square Method, and the value of a, b can be gained as shown in Table 6. Figure 14(b) shows the numerical results and the fitting curve originated from fitting formula when B = 0.3. It can be concluded that the errors between the numerical results and fitting curve are very small, which means that the assumption of H_2 RBL is acceptable.

Table 6: Fitting formulas about H_2 RBL and gas flow velocity

B	RBL fitting formula/100%
0.2	Y = -0.03035 lgV + 0.77462
0.3	Y = -0.03147 lgV + 0.78372
0.4	Y = -0.03205 lgV + 0.78404
0.5	Y = -0.03212 lgV + 0.78627
0.6	Y = -0.03125 lgV + 0.77937
0.7	Y = -0.03080 lgV + 0.77650

Figure 15 shows the relationship between a or b and B. Table 6 and Figure 15 show that as the blockage ratio increases, the slope a will increase first and decrease later, while the intercept b will decrease first and increase later. The relationship between a or b and B is approximately a quadratic function, so the formula for a, b and B can be assumed as follows:

$$a = m_0 + m_1 B + m_2 B^2,$$

$$b = n_0 + n_1 B + n_2 B^2.$$

$$(4.3)$$

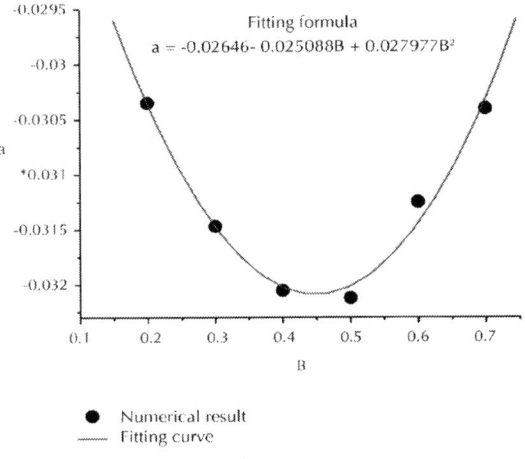

(a)

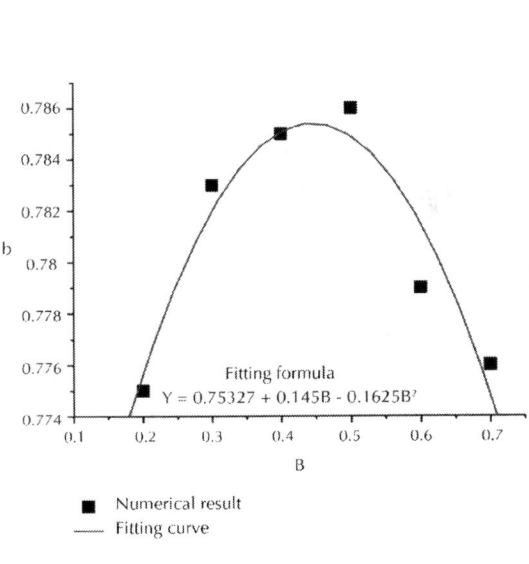

(b)

Figure 15: Slope and intercept of fitting function of H₂ RBL versus blockage ratio . (a) Slope fitting curve, (b) Intercept fitting curve.

Fitting the curve with a quadratic function by Least Square Method, then a formula for a, b and B can be summarized as follows (V ≤ 50 m/s):

$$a = -0.02646 - 0.025088B + 0.027977B^2,$$

$$b = 0.75327 + 0.145B - 0.1625B^2.$$

(4.4)

Figure 16 shows that the distance from the bluff-body to the recirculation center will increase as the blockage ratio increases, and so does the recirculation length. But while the blockage ratio remains unchanged, the position and length of the recirculation region is unchanged, no matter how large the flow velocity is. It means that blockage ratio significantly influences the recirculation region length. Wright [4] found that when the situation is approaching to blow off, the residual flame occupies just the recirculation zone region, and the recirculation-zone length remains unchanged. Only the blockage has a strong influence on recirculation-zone length. For a given blockage ration, the wake with after the bluff-body is virtually constant, independent of mixture ratio and flow speed. So, people cannot judge the blowout limit just by recirculation length.

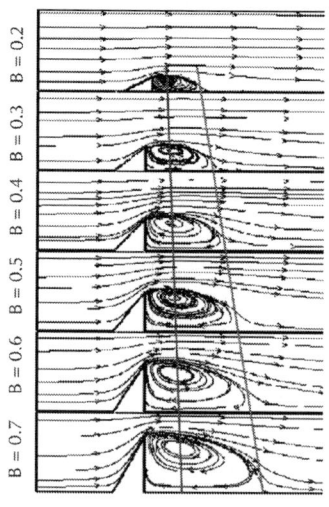

(a)

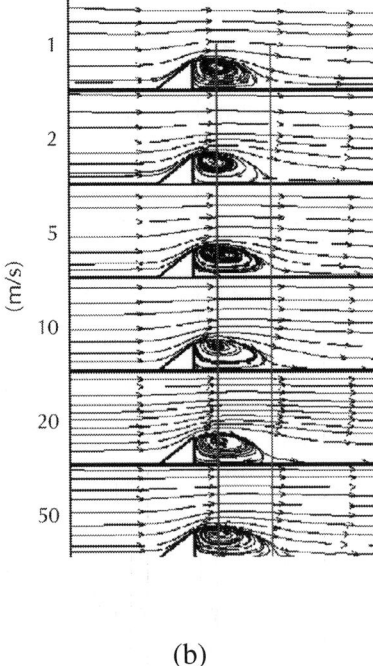

(b)

Figure 16: Streamlines of recirculation zone. (a) Recirculation versus block-age ratio and (b) Recirculation versus flow velocity.

Ignition Process Analysis

The simulation of ignition process is done in the condition of d = 30 mm, V= 2 m/s, and equivalence ratio ϕ = 0.5. A small zone is patched with high temperature, so the burning will begin. With the development of ignition process, the flame flied becomes stable after t = 200 ms, and the calculation of ignition process continues to t = 500 ms for ensuring the stability of the flow field. When t = 500 ms, reduce the H₂ volume concentration to 0.095 (ϕ = 0.25, below the H₂ RBL) to simulate the extinction process.

Start ignition: t= 0 ms (equivalence ratio is ϕ =0.5);

Ignition process: t = 0–500 ms;

Extinction process: t = 500–1400 ms (equivalence ratio is ϕ = 0.25).

First half parts of Figure 17 (before the vertical dash line) and Figure

18(a) show the species mass fraction and temperature distribution of section x=40mm in ignition process. When t=100ms, the average temperature of section x=40 will increase sharply, so does the mass fraction of H_2O, OH, O, H, while the mass fraction of hydrogen and oxygen will decrease greatly, these phenomena mean that the flame has been ignited at this time. In the period of 100ms to 200ms, the average temperature of section x=40mm will decrease suddenly, and meanwhile, the mass fraction of H_2O, OH, O, and H will also decrease, and then remain unchanged. The concentration of H_2 and O_2 would rebound after t=100 ms, and this means that the flame is not stable until t=200ms, so it can be seen that the ignition sequence is not successful until t=200ms.

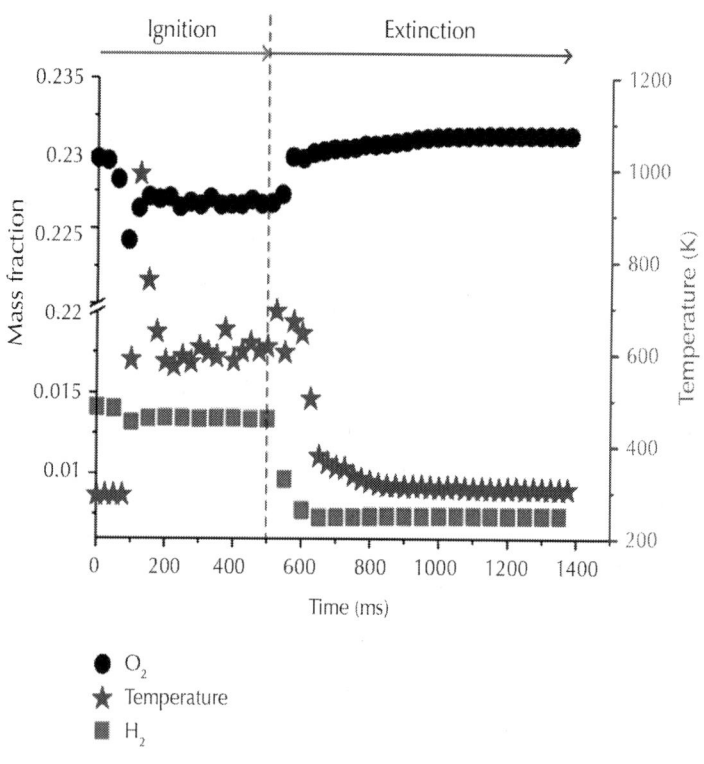

(a)

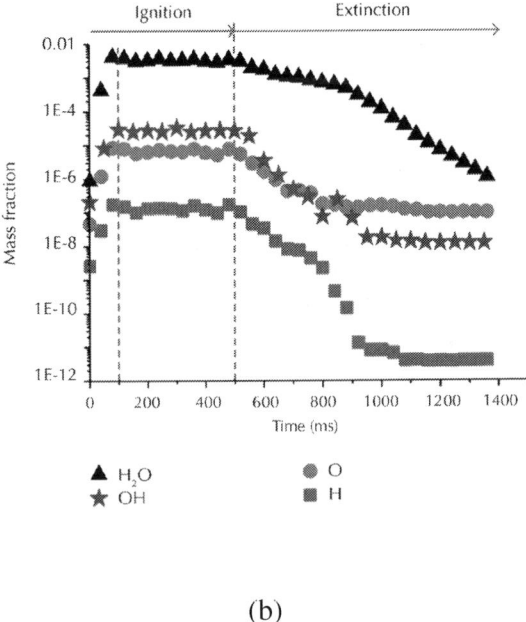

(b)

Figure 17: Profiles of species average mass fraction and average temperature on section mm versus time from ignition (0–500ms) to extinction (500–1400ms) **(a)** O$_2$, H$_2$, Temperature and **(b)** H$_2$O, OH, H.

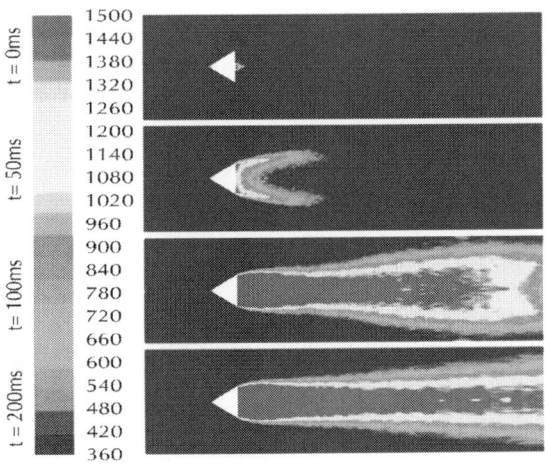

(a)

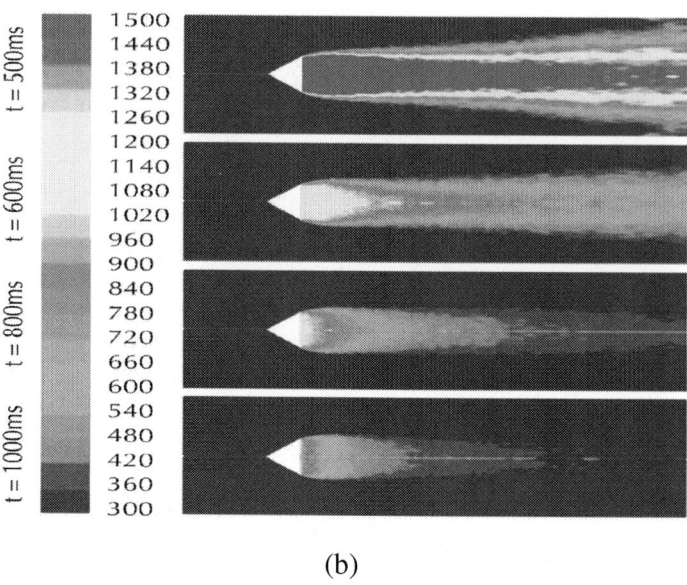

(b)

Figure 18: Temperature distribution from ignition process to extinction process (a) Ignition process and (b) Extinction process.

Figure 18(a) shows the temperature field of whole ignition process. There is a fluctuation of the flame during ignition process as shown in Figures 17 and 18(a) This is mainly because when the burning has just taken place first in CRZ, a large amount of combustion and intermediate products generated with heat releasing from chemical reaction, but the heat and the intermediate products cannot spread to the main stream immediately, and they would accumulate rapidly at this time. While t=100ms, the remainder reactants and the accumulated intermediate products have almost been consumed completely. In order to maintain burning, fresh reactants are required, and at this moment, the heat which spread to the main stream from recirculation zone is more than the heat released from burning, so the temperature and the chemical reaction in the recirculation zone will decline. When t=200ms, the heat released from fresh burning in the recirculation zone is adequate to compensate for the heat taken away by the main stream, and up to now, the combustion is steady.

It can be concluded that a successful ignition sequence in a bluff-body burner requires three phases: (1) the startup of ignition in the recirculation zone; (2) the energy accumulation in the recirculation zone; (3) the flame propagation from the recirculation zone to the main stream.

Extinction Process Analysis

When t=500ms, in order to investigate the flame extinction process, hydrogen concentration was reduced to 0.095 (lower than the flammable limit), and the extinction will happen.

Latter half parts of Figure 17 (after the vertical dash line) show that it takes about 100ms for the whole combustion field to reduce H$_2$ mass fraction to the new value 0.095. It can be seen from Figure 17 that the whole extinction process takes about 600ms, which is longer than ignition. When t=600ms, the fresh mixture has reached the CRZ completely, and the gas concentration in the recirculation is below the flammable limit, but the flame does not extinguish immediately, this is because the energy dissipation from the recirculation zone to the main stream is slow, so the temperature in the recirculation zone is still high enough to maintain the burning for a while, and the flame will not extinguish until the energy in the recirculation totally diffuse to the main flow after the temperature in the recirculation zone reduces to the extinguish level, and the flame goes out totally as Figure 18(b) shows.

During t= 600–800ms, the temperature attenuation rate would reach a new value, meaning that the burning is still going on below H$_2$ RBL, and the flame length would reduce gradually. When t= 800–1400ms, the average temperature will slowly decrease to the cold field level, this process is the burning of remaining gas in the recirculation. So far, the flame has extinguished completely. Figure 18(b) shows that the flame will take an "M" shape with reaction fronts inside the CRZ near the blow off condition. This flame shape is in well agreement with that gained by Dawson et al. [5] in their experiment measurement.

In a word, C-PDF model is accurate enough to capture the flame extinction. In terms of control of marine power, it can be concluded that the flame will extinguish as soon as the average temperature is lower than that at 800ms. Feedback should be provided to fuel and air control system promptly to regulate fuel supply in order to avoid extinction.

So, a complete flame extinction process requires three phases: (1) the sudden decline of temperature in the burner because of the decline of fuel concentration; (2) the energy dissipation from the recirculation zone to the main stream; (3) the flame complete extinction.

Sensitivity Analysis for Chemistry Reaction

The sensitivity analysis is a powerful tool in interpreting the results of computational simulations, and it can be used to research the influence of temperature, species concentration, and equivalence ratio on each elementary reaction [18, 19]. Rate-of-production analysis provides complementary information on the direct contributions of individual reactions to species net production rates.

To investigate the contribution of each elementary reaction to H_2 burning, the software CHEMKIN [18] was used to analyze the first-order sensitivity coefficient of temperature and species according to the H_2 18 steps reaction mentioned earlier in the paper. Finally, make use of sensitivity coefficient to investigate the ignition and extinction process. CHEMKIN assumes a variable Z as

$$\frac{dZ}{dt} = F\left(Z, t, \vec{a}\right),$$

(4.5)

where $Z = (Y_1, Y_2,...,Y_k)$ T are standards for mass fraction of each species, $\bar{a} = (A_1, A_2, ...,A_N)$-preexponential factor of each species, and the first-order sensitivity coefficient is defined as

$$w_{l,i} = \frac{\partial Z}{\partial a_i},$$

$$\frac{dw_{l,i}}{dt} = \frac{\partial F_l}{\partial Z} \cdot w_{l,i} + \frac{\partial F_l}{\partial a_i}.$$

(4.6)

For heat-of-formation sensitivity, $\bar{a}$ represents the vector of heats of formation for all the species in the system. The change of $\bar{a}$ will bring in the species concentration variety. The bigger sensitivity coefficient means the more significant influence caused by $\bar{a}$.

The equivalence ratio for calculation case in Figure 19 to Figure 23 was 1. Figure 19 shows that reaction R2 possesses the largest positive temperature sensitivity coefficient other than all the other elementary reactions. The sensitivity coefficient reaches the peak value at 1550K. It indicates that at T= 1550K, a little temperature variation will induce a large chemistry reaction rate change for R2 and R9. So, temperature has a very important influence on reactions R2 and R9. According to formula (4.8), R2 is an exothermic reaction, so in ignition process, if we can improve the mixture temperature to 1550K by a spark in a short time, the exothermic reaction R2 will occur immediately, and the heat will be released from R2 rapidly. So R2 is very important to ignition process. In contrast, reaction R9 possesses the largest negative temperature sensitivity coefficient in all the elementary reactions. It means that increasing its rate will lead to a lower temperature. So, reactions R2 and R9 dominate the ignition and extinction processes. R9 ought to be the first reaction taking place in ignition process, and R2 ought to be the last one consider

$$OH + O \Longrightarrow O_2 + H$$

(4.7)

$$\Delta h = [h(O_2) + h(H)] - [h(O) + h(OH)] = -70(kJ/mole)$$

(4.8)

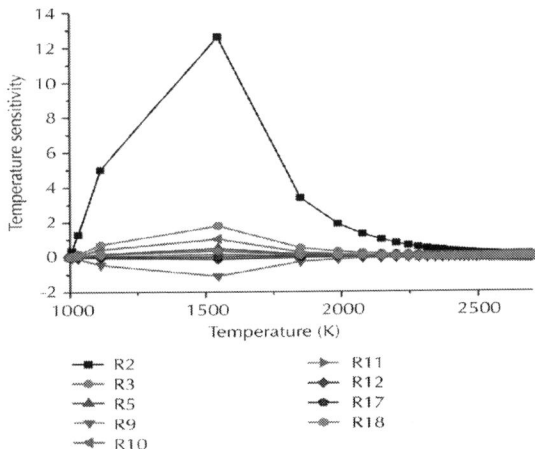

Figure 19: Temperature sensitivity for each reaction.

To investigate the influence of species concentration on chemistry reaction, the species sensitivity of intermediate species (H, O, and OH), reactants, and production were carried out. Figure 20 shows that to R2, the species sensitivity (O, OH) would decrease first and rise up later as temperature increases. When the temperature reaches 1550K, the species sensitivity reaches the lowest value. It means that the species change rate reaches the maximum value. The H sensitivity of R9 reaches to the lowest value at 1120K which is earlier than species O and OH, which indicates that reaction R9 ought to be taking place earlier than R2. Because in ignition process the fresh mixture must obtain heat from igniter source, the ignition will not be successful until the flame core accumulates enough heat. Figures 21 and 22 show that before 1550K, the reactant (H_2, O_2) mole fraction will decline rapidly and the intermediate species and production concentration will rise greatly. It means that the endothermic reactions R1, R3, R5, and R9 take place immediately when the temperature rises from 1000K to 1550K. When the temperature exceeds 1550K, the decreasing rate of species H_2 and O_2 is lower than that before 1550K, which indicates that the flame was not ignited until T=1550K.

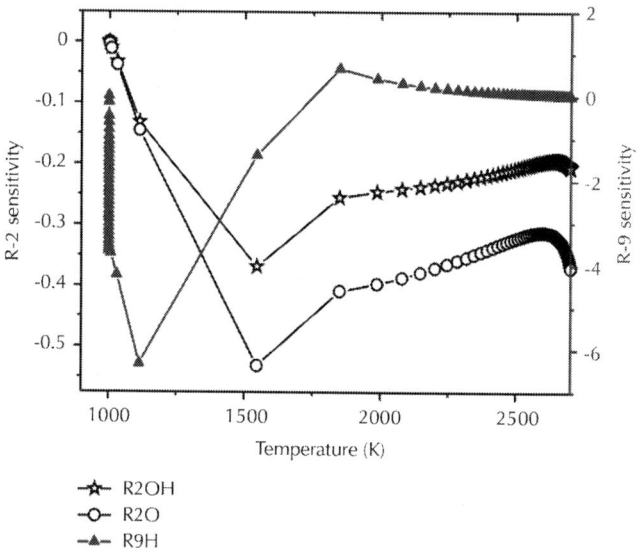

Figure 20: Temperature sensitivity for intermediate species of R2 and R9.

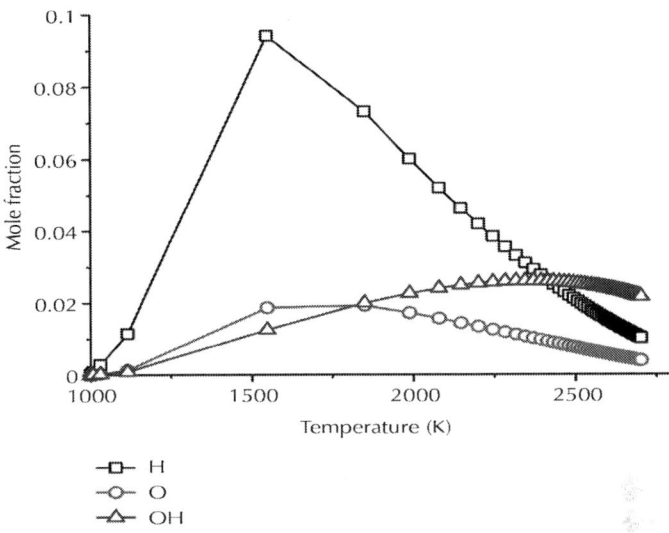

Figure 21: Profiles of intermediate species versus temperature.

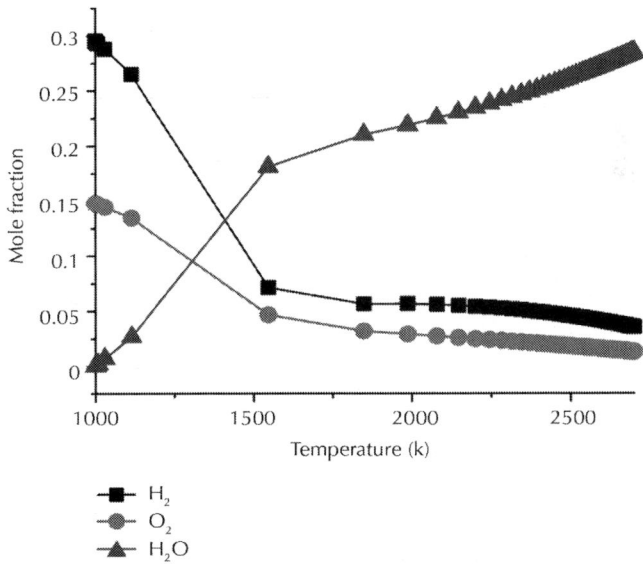

Figure 22: Profiles of reactants and production versus temperature.

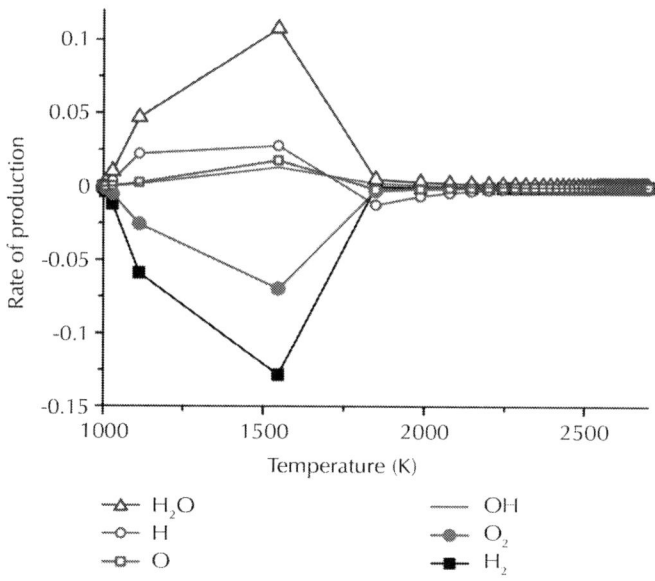

Figure 23: Rate of production versus temperature.

Figure 23 shows that the consumed rate of the reactant (O_2, H_2) increases rapidly before 1550K and changes a little when the temperature reaches to 1850K then remains unchanged. The same is done to intermediate species (O, OH, H) and production. It indicates that the ignition is successful after K, and the flame comes into being homeostasis after 1850K.

Figure 24 shows the relationship between temperature sensitivity and equivalence ratio (Φ = 0.1~10.0). It can be seen that the temperature sensitivity of reaction R2 will rise first and decrease later as equivalence ratio increases. The temperature sensitivity reaches to the peak value when equivalence ratio rises to 2. When equivalence ratio is lower than 0.1 or higher than 10, the chemistry reaction will not take place because it has exceeded the combustibility limit. Figure 25 shows that the temperature sensitivity of R9 will decrease first and increase later with the increasing equivalence ratio, and it will reach the lowest value when equivalence ratio reaches 1. It means that when equivalence ratio is 1, the mixture is most ignitable. In a word, people can control the chemistry reaction process, flame temperature, or ignition process by adjusting the mixture equivalence ratio.

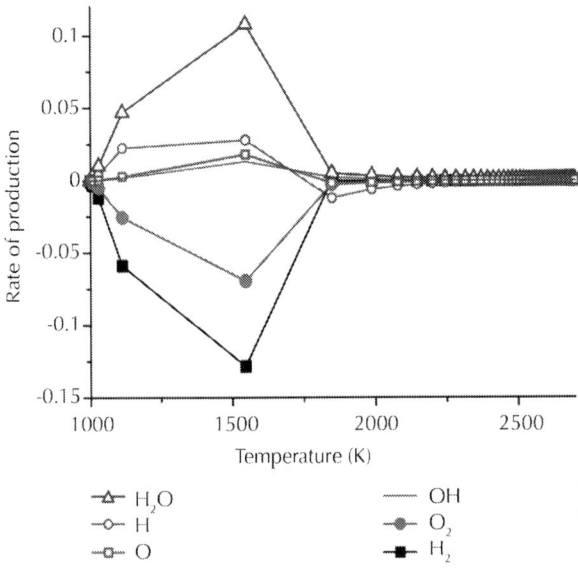

Figure 24: Temperature sensitivity coefficient of R2 versus equivalence ratio.

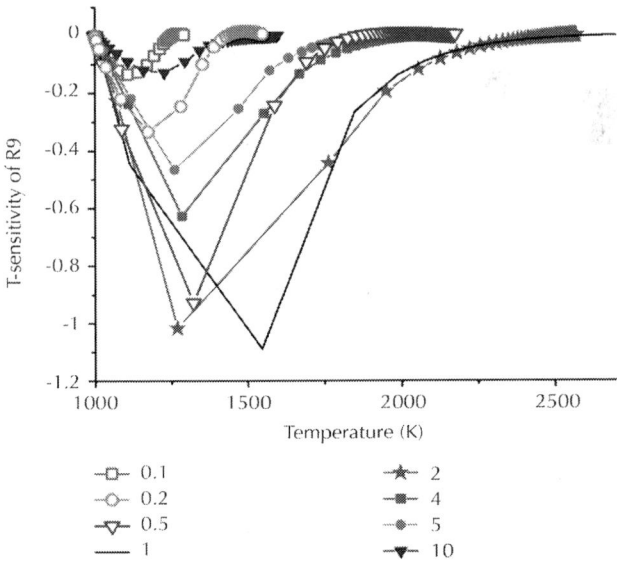

Figure 25: Temperature sensitivity coefficient of R9 versus equivalence ratio.

CONCLUSIONS

The numerical simulation on H_2 premixed flame in a bluff-body burner has been carried out. The H_2 flame ignition and extinction process is analyzed, and a function formula is summarized for H_2 RBL. The results showed that k-epsilon-C-PDF model is a reasonable method to capture H_2 RBL. There should be an optimal blockage ratio, which can stabilize the flame best. The flame will take an "M" shape with reaction fronts inside the CRZ near the blow off condition. This research can provide theoretical instruction for bluff-body burner design, gas flow velocity control, fuel concentration matching, and the flame stability research. To systematically analyze the role of each elementary chemistry reaction taking place in the global combustion, CHEMKIN software was adopted to investigate the sensitivity of each elementary reaction. Other conclusions are as follows.

- When the blockage ratio remains unchanged, H_2 rich blowout limit is gas flow velocity's logarithmic function.

- When the gas flow velocity remains unchanged, H_2 rich blowout limit is blockage ratio's quadratic function.

- The fitting formula of H_2 rich blowout limit is $Y = algV + b$ ($V \leq$ 50 m/s), where, B is blockage ratio, $a = -0.02646 - 0.025088B + 0.027977B^2$, and $b = 0.75327 + 0.145B - 0.1625B^2$.

- A complete extinction process in a bluff-body burner requires three phases, the suddenly decline of the temperature in the main stream, the energy dissipation from the recirculation zone to the main stream, and the flame complete extinction.

- Reactions R2 and R9 possess the largest positive and negative temperature sensitivity. Increasing the rate of R2 will lead to a higher temperature, and increasing the rate of R9 will lead to a lower temperature. When equivalence ratio is 1, the mixture is most ignitable. The critical ignition temperature is 1550K. Temperature has a very important influence on reactions R2 and R9.

REFERENCES

1.   S. J. Shanbhogue, S. Husain, and T. Lieuwen, "Lean blowoff of bluff body stabilized flames: scaling and dynamics," Progress in Energy and Combustion Science, vol. 35, no. 1, pp. 98–120, 2009.

2.   P. Eriksson, "The Zimont TFC model applied to premixed bluff body stabilized combustion using four different rans turbulence models," in Proceedings of the Turbo Expo: Power for Land, Sea, and Air Conference, pp. 353–361, Montreal, Canada, May 2007.

3.   S. M. Frolov, V. Y. Basevich, and A. A. Belyaev, "Mechanism of turbulent flame stabilization on a bluff body," Chemical Physics Reports, vol. 18, no. 8, pp. 1495–1516, 2000. ·

4.   F. H. Wright, "Bluff-body flame stabilization: blockage effects," Combustion and Flame, vol. 3, pp. 319–337, 1959.

5.   J. R. Dawson, R. L. Gordon, J. Kariuki, E. Mastorakos, A. R. Masri, and M. Juddoo, "Visualization of blow-off events in bluff-body stabilized turbulent premixed flames," Proceedings of the Combustion Institute, vol. 33, no. 1, pp. 1559–1566, 2011.

6.   P. Griebel, E. Boschek, and P. Jansohn, "Lean blowout limits and NOx emissions of turbulent, lean premixed, hydrogen-enriched methane/air flames at high pressure," Journal of Engineering for Gas Turbines and Power, vol. 129, no. 2, pp. 404–410, 2007.

7.   R. W. Schefer, "Hydrogen enrichment for improved lean flame stability," International Journal of Hydrogen Energy, vol. 28, no. 10, pp. 1131–1141, 2003. ·

8.   R. S. Barlow, J. H. Frank, A. N. Karpetis, and J. Y. Chen, "Piloted methane/air jet flames: transport effects and aspects of scalar structure," Combustion and Flame, vol. 143, no. 4, pp. 433–449, 2005. ·

9.   E. Giacomazzi, V. Battaglia, and C. Bruno, "The coupling of turbulence and chemistry in a premixed bluff-body flame as studied by LES," Combustion and Flame, vol. 138, no. 4, pp. 320–335, 2004.

10. C. X. Lin and R. J. Holder, "Reacting turbulent flow and thermal field in a channel with inclined bluff body flame holders," Journal of Heat Transfer, vol. 132, no. 9, pp. 1–11, 2010.

11.   W. W. Kim, J. J. Lienau, P. R. Van Slooten, M. B. Colket, R. E. Malecki, and S. Syed, "Towards modeling lean blow out in gas turbine flameholder applications," Journal of Engineering for Gas Turbines and Power, vol. 128, no. 1, pp. 40–48, 2006.

12.   W. P. Jones and V. N. Prasad, "Large eddy simulation of the sandia flame series (D-F) using the Eulerian stochastic field method," Combustion and Flame, vol. 157, no. 9, pp. 1621–1636, 2010.

13.   Ansys Fluent 12.0 Documentation.

14.   F. Bisetti and J. Y. Chen, LES of Sandia Flame D with Eulerian PDF and Finite-Rate Chemistry, Combustion Modeling, Combustion Processes Laboratories, Berkeley, Calif, USA, 2005.

15.   R. P. Lindstedt, S. A. Louloudi, and E. M. Váos, "Joint scalar probability density function modeling of pollutant formation in piloted turbulent jet diffusion flames with comprehensive chemistry,"Proceedings of the Combustion Institute, vol. 28, no. 1, pp. 149–156, 2000.

16.   S. M. Correa and S. B. Pope, "Comparison of a monte carlo PDF/ finite-volume mean flow model with bluff-body raman dsata," in Proceedings of the 24th Symposium International on Combustion, pp. 279–285, The Combustion Institute, 1992.

17.   S. M. S. El-feky and A. Penninger, "Study of flammability lean limit for a bluff body stabilized flame,"Periodica Polytechnica, Mechanical Engineering, vol. 38, no. 1, pp. 33–45, 1994.

18.   Reaction Design, "Theory manual," CHEMKIN Release 4. 1. 1.

19.   Y. T. Liang and W. Zeng, "Kinetic simulation of gas explosion in constant volume bomb," Journal of Combustion Science and Technology, vol. 16, no. 4, pp. 375–381, 2010.

Simulation and Assessment of SO$_2$ Toxic Environment after Ignition of Uncontrolled Sour Gas Flow of Well Blowout in Hills

Yuan Zhu and Guo-ming Chen

Department of Safety Engineering, China University of Petroleum, Dongying 257061, China

ABSTRACT

To study the sulfur dioxide (SO$_2$) toxic environment after the ignition of uncontrolled sour gas flow of well blowout, we propose an integrated model to simulate the accident scenario and assess the consequences of SO$_2$ poisoning. The accident simulation is carried out based on computational fluid dynamics (CFD), which is composed of well blowout dynamics, combustion of sour gas, and products dispersion. Furthermore, detailed complex terrains are built and boundary layer flows are simulated according to Pasquill stability classes. Then based on the estimated exposure dose derived from the toxic dose–response relationship, quantitative assessment is carried out by using equivalent emergency response planning guideline (ERPG) concentration. In this case study, the contaminated areas are graded into three levels, and the areas, maximal influence distances, and main trajectories are predicted. We show that wind drives the contamination and its distribution to spread downwind, and terrains change the distribution shape through spatial aggregation and obstacles. As a result, the most dangerous regions are the downwind areas, the foot of the slopes, and depression areas such as valleys. These cause unfavorable influences on emergency response for accident control and public evacuation. In addition, the effectiveness of controlling the number of deaths by employing ignition is verified in theory. Based on the assessment results, we propose some suggestions for risk assessment, emergency response and accident decision making.

INTRODUCTION

For sour gas well blowout, which is one of the most serious accidents resulting from gas field exploitation, ignition of uncontrolled flow is recommended or constrained by various standards, laws, or directives [1], [2] and [3]. In China, after the disaster of the '12.23' Kaixian blowout accident in which 243 people died mostly due to a too-long-delayed ignition decision [4], immediate ignition under extreme conditions was emphasized in drilling safety in such gas fields. For example, the new enterprise standard focusing on the highly sour gas field exploitation in the northeast of Sichuan demands that out-of-control well blowouts must be ignited within 5 min [2]. To comply with the supervision requirements, ignition is probably what will be faced in the event of an accidental sour gas well blowout. During the combustion of sour gas, huge amounts of sulfur dioxide (SO_2) which is an irritation of the upper respiratory tract and eyes will be produced and can cause serious injury to people [1]. Consequently, there is a potential threat to the general public surrounding the well of exposure to the toxic environment formed by the dispersed SO_2.

Detailed safety analysis of sour gas well blowout has been carried out because of the serious consequences. Research on the Kaixian disaster, in particular, has extended our knowledge in this field. Li et al. elaborated the basic information of the accident and made a systematical analysis [4]. Wellhead jet dynamics used for further research was modeled according to well-bore inflow [5]. Movement of hydrogen sulfide in complex terrains and its influence have been simulated based on computational fluid dynamics (CFD) and validated by accident investigation [6] and [7]. However, these works are still lacking in regard to ignition and SO$_2$ poisoning.

Therefore, analysis of the ignition process and quantitative assessment of the risk of SO$_2$ are very much needed. In this field, the Energy Resources Conservation Board (ERCB) releases dedicated software, ERCBH$_2$S that includes an assessment model for SO$_2$. Due to the method of parallel airflow modeling, this model is more suitable for flat terrains [8]. However, as most sour gas fields are in the hill regions of the northeast of Sichuan in China, the effects of complex terrains on gas dispersion should be included. Although much research on the atmospheric flow over hills has been carried out based on field measurement, laboratory experiments, and numerical methods [6], [7], [9], [10] and [11], reactive sour gas transportation is not mentioned. In these works, CFD simulation, with the advantages of low cost, high efficiency, and powerful modeling capabilities, has been widely adopted [6], [7], [10] and [11]. The results of the comparison between CFD prediction and experiments show that this method is quite useful when modeling plume dispersion on complex topography. For reactive pollutant transportation, the CFD method has been applied to simulate the fire-induced plume dispersion in street canyons [12]. As the governing equations are the same for the reactive dispersion, the model used in street canyons could be transferred to complex terrains. And this method has been used by ENI Exploration and Production to do a complete risk analysis of the ignited jet associated with the blowout, in which a series of combustion simulations is carried out and the far-field impact about soot is assessed [13] and [14].

In order to achieve systemic and comprehensive analysis, CFD is adopted to simulate the accident by using FLUENT software [15]. Then the assessment of toxic environment is made based on predicted concentration and distribution of SO$_2$.

THEORETICAL MODEL

The accident process is divided into three consecutive stages: wellhead jet, sour gas combustion and products dispersion. As a whole, in order to achieve simulation in a consistent model, the ideal gas law for an incompressible flow is applied and gas density is computed as:

$$\rho = \frac{P_a M}{RT}$$

(1)

This is a common assumption in CFD simulation on atmospheric flow [6] and [7]. For combustion, as far-field assessment on SO_2 is the main concern in this study, the pressure expansion due to combustion is ignored so that the rise of plume is underestimated, which produces more conservative results for the safety assessment. This assumption is not suitable for wellhead jets. The method used is described below.

Well Blowout Dynamics

To evaluate the worst-case accident scenario, wellhead absolute open flow is commonly supposed to analyse sour gas well blowout [5] and [8]. Though detailed modeling of well-bore inflow is more proper to assess the surface release conditions, it is too complex to achieve and beyond the scope of this simulation. Whereas assuming the well bottom pressure is high enough so that the gas flow on the wellhead can reach choked condition, the initial condition for the wellhead mass flux is given by:

$$m_E = V_E \rho_E A$$

(2)

Based on Eq. (2), a balance between the mass flux, choked condition and casing or tubing inside diameter at well head can be achieved [8]. It is much easier to apply this method to do assessment, especially for a planned or under construction well, than to calculate the well-bore inflow which needs detailed information about the reservoir and casing program.

Then the release expands to atmospheric pressure and loses momentum due to changes in direction or impingement with the ground or structures without air entrainment. At the end of this process, the gas equation of state is applied as the pressure is equal to atmospheric pressure, and the well blowout is introduced to the CFD model as velocity-inlet boundary defined by [8]:

$$V = \left(V_E + \frac{P_E - P_a}{\rho_E V_E}\right)\left(\frac{1 - C_D}{1 + C_D}\right)$$

(3)

Detailed calculation is based on ERCBH$_2$S, which accounts for real gas property and conservation of mass, momentum, and energy.

Combustion Simulation

The mixture fraction-based combustion model is assumed for the ignition of sour gas, which is more simple and efficient in fire simulation compared with the finite-rate reaction [16]. As the calculation is done in large space, the reduction in computing resource and time is important.

The basic assumption is under a certain set of simplifications, the instantaneous thermochemical state of the fluid can be related to a conserved scalar quantity known as the mixture fraction f [17]. For every element involved in the combustion, f is defined as:

$$f = \frac{Z - Z_{ox}}{Z_{fuel} - Z_{ox}}$$

(4)

All the chemical reactions between sour gas (mostly methane, hydrogen sulfide and carbon dioxide) and atmosphere (mostly nitrogen and oxygen) are simplified into one single reaction without phase change:

$$CH_4 + 2H_2S + 5O_2 \rightarrow CO_2 + 4H_2O + 2SO_2$$

By introducing f, the chemistry is simplified into a mixture question, and the solving of nonlinear finite-rate chemistry is avoided. Under the assumption of chemical equilibrium, thermochemical scalars (species fractions, density, and temperature) are related to f. For a single mixture fraction, non-adiabatic system, the instantaneous values of thermochemical scalars are parameterized as:

$$\phi_i = \phi_i(f, H)$$

$$(5)$$

Regarding chemical nonequilibrium, the rich flammability limit (RFL) method of equilibrium chemistry model is used, in which fuel-rich regions are modeled as a mixed but unburnt mixture of pure fuel and a leaner equilibrium burnt mixture. The RFL of the fuel stream is estimated as 110–150% of the stoichiometric mixture fraction.

Gas Dispersion Model

Large eddy simulation (LES) associated with the mixture fraction transport equation is used as the theoretical basis for gas dispersion model, which is capable of simulating jet fire, reactive pollutant dispersion, and boundary flow in complex terrains. With fine grid resolution, a close agreement with experimental observations in jet fire simulation could be provided by LES [17]. Hu et al. studied the fire-induced buoyancy-driven plume in and above an idealized street canyon by using LES [12]. By comparing wind-tunnel observations with the predictions of various turbulence closure models, LES was shown to be capable of reproducing sensible results for flow over rough hills by Allen and Brown [18].

In LES, flow features that are larger than the grid size are resolved directly, whereas flow structures that are smaller than the grid size are modeled using a subgrid scale model. So a better balance between resource consumption and result accuracy is achieved compared with two other CFD turbulence models: direct numerical simulation (DNS) and Reynolds averaging Navier–Stokes equation (RANS). LES is more efficient in computation than DNS, and exhibits better agreement with the experimental results than RANS. Simulations of the neutrally stratified flow over Askervein Hill showed that LES provided an acceptable solution for the mean-velocity field and better predictions of the turbulent kinetic energy than RANS [11]. For transient mixing processes and the transient structure of turbulent fields, LES would be a better choice compared to RANS, as suggested by Li et al. [19]. The method used here is described below.

The equations for mass and momentum are as follows [17]:

$$\frac{\partial \overline{\rho}}{\partial t} + \frac{\partial}{\partial x_i}(\overline{\rho}\widetilde{u_i}) = 0$$

(6)

$$\frac{\partial}{\partial t}(\overline{\rho}\widetilde{u_i}) + \frac{\partial}{\partial x_j}(\overline{\rho}\widetilde{u_i}\widetilde{u_j}) = -\frac{\partial \overline{p}}{\partial x_i} + \frac{\partial}{\partial x_j}\left[(\mu + \mu_t)\left(\frac{\partial \widetilde{u_i}}{\partial x_j} + \frac{\partial \widetilde{u_j}}{\partial x_i}\right)\right]$$

$$+ \overline{\rho} g_i$$

(7)

Where μ_t is calculated by the Smagorinsky–Lilly model.

Under the assumption that the diffusivity of all species is equal and gas dispersion is dominated by turbulent convection, the species transport equations are replaced by a single mixture fraction equation:

$$\frac{\partial}{\partial t}(\overline{\rho}\overline{f}) + \frac{\partial}{\partial x_i}(\overline{\rho}\widetilde{u_i}\overline{f}) = \frac{\partial}{\partial x_i}\left(\frac{\mu_t}{\sigma_t}\frac{\partial \overline{f}}{\partial x_i}\right) + S$$

(8)

The turbulence–chemistry interactions are modeled by an assumed-shape probability density function approach [20]. For the relation between mean thermochemical scalars and their instantaneous values, assuming that the enthalpy fluctuations are independent of the enthalpy level, the equation is:

$$\overline{\phi_i} = \int_0^1 \phi_i(f, \overline{H})p(f)df$$

(9)

Where $p(f)$ is chosen as the double delta function, and is given by:

$$p(f) = \begin{cases} 0.5, & f = \overline{f} \pm \sqrt{\overline{f'^2}} \\ 0, & \text{elsewhere} \end{cases}$$

(10)

Where $f' = f - \overline{f}$ and the mixture fraction variance is modeled in LES as:

$$\overline{f'^2} = C_v L_s^2 \left|\nabla \overline{f}\right|^2$$

(11)

The transport equation for total enthalpy (sum of sensible and formation enthalpies) is:

$$\frac{\partial}{\partial t}(\bar{\rho}\bar{H}) + \frac{\partial}{\partial x_i}(\bar{\rho}\tilde{u}_i\bar{H}) = \frac{\partial}{\partial x_i}\left(\frac{k_t}{c_p}\frac{\partial \bar{H}}{\partial x_i}\right) - S_{rad} + S_h$$

(12)

With the assumption of incompressible flow, the pressure work, kinetic energy, and viscous dissipation terms are not included [15]. The first term on the right-hand stands for the heat transfer due to conduction and species diffusion.

To fully account for the radiation, the radiative transfer equation should be solved which needs large computation [15]. As the hazard deriving from the radiation is not the concern in this study, a simple method is proposed to treat the heat loss due to radiation.

The total heat of combustion converted to radiation is calculated as:

$$Q_{rad} = \eta m_E H_c$$

(13)

Where   is conversion factor for methane [21] which is the largest component of sour gas.

Then the S_{rad} is defined as:

$$S_{rad} = \frac{Q_{rad}}{V_f}$$

(14)

Where V_f could be determined from [22]:

$$L_f = 0.00326(m_E H_c)^{0.478}$$

$$R_f = 0.29s_f\left[\log_{10}\left(\frac{L_f}{s_f}\right)\right]^{0.5}$$

(15)

And it is added to the computational domain with the temperature higher than 773.15 K. The advantage of this method is that radiation is solved directly by means of energy absorption and no additional iterations are needed. Besides, the energy transferred to atmospheric environment is discounted so that the transport of product gas surrounding the flame is reduced to get a conservative estimate of the gas dispersion.

CFD CALCULATION

Complex Terrain Modeling

A simple and practical method is proposed to reconstruct the terrains in the CFD geometrical model, which is based on SRTM 90-m digital elevation data for the entire world. First, natural relief is extended to a ramp zone around the boundaries. Next, spline interpolation is applied along each longitude and latitude to get continuous edges. Then, a four-sided surface standing for the real terrains is created from the two sets of parallel edges. This method is easy to be applied to other areas without the support of special software or tools, and the stiff curvature of reconstructed terrains is avoided by interpolation under the assumption of continuous terrains.

An example well from the Puguang highly sour gas field in the northeast of Sichuan and its geometrical model are showed in Fig. 1. The size of the computational domain is 3960 m × 3960 m × 1600 m, with the well set in the middle and north pointing at the +Y-axis. In the actual environment around the well, elevations increase toward the west and the maximum difference in altitude is about 300 m. In the east and south, the well is enclosed by a rivulet flowing from west to north. And from the lowest valley in the north to the well is about 180 m in elevation.

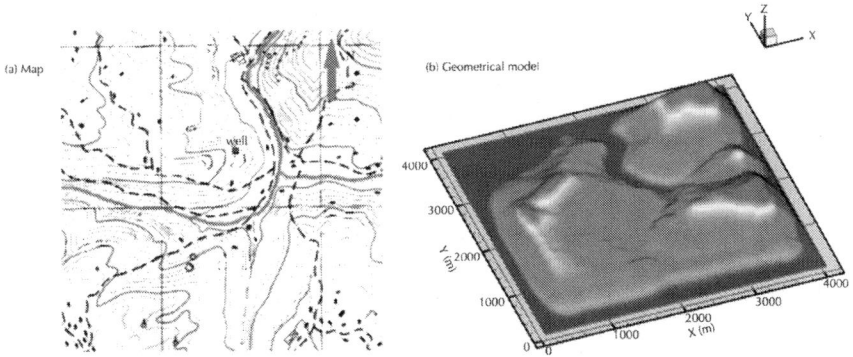

Figure 1: The location of the example well. (a) Map, and (b) Geometrical model.

Dispersion Conditions

Dispersion conditions have great influence on the consequences of accidents. In the atmospheric environment of low wind speed, temperature inversion, and stable atmosphere, the worst-case situation would be faced, caused by weak diffusion and sustained steady accumulation of SO_2. In order to reflect the atmospheric phenomena in the CFD model, the boundary layer flows are simulated by using the Pasquill atmospheric stability classes, and the corresponding profiles of wind speed and temperature along the altitude.

Class F is used to do conservative analysis. Then the empirical exponential law for wind speed profile is written as follows [23]:

$$\frac{v_1}{v_2} = \left(\frac{z_1}{z_2}\right)^{0.3}$$

(16)

For standard wind speed, according to environmental statistics on site, the mean wind speed is 1.0–1.8 m/s measured 10 m from the ground in all directions. The prevailing wind directions are calm winds, with the frequency of 37% and northeast 24% all year. The assessment is conducted with calm wind, and eight wind directions (E, NE, N, NW, W, SW, S and SE) with an annual average wind velocity of 1.3 m/s. The vertical temperature gradient is 4 K/100 m [23] and an annual average temperature of 289.95 K is used.

Numerical Method

The three-dimensional computational domain is totally mapped with regular hexahedral elements after quadrilateral mesh generation for all boundaries and every part of the terrains. Vertical grid refinement is applied to side boundaries and core computational domain, and is shown in Fig. 2 on side plane and the slice.

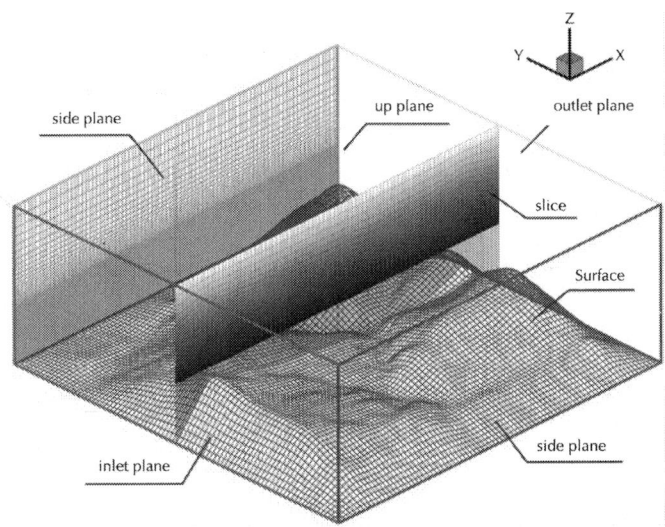

Figure 2: Illustration of mesh and boundaries.

In consideration of practical application, the domain is divided into 135 (X) × 138 (Y) × 116 (Z). The average grid spacing in the horizontal directions (X and Y) is between 12.5 and 100 m in the gas dispersion modeling over complex terrains of large scale [11], [24] and [25]. The value we selected is 30 m. In the vertical direction (Z), the grid spacing from the ground to the top of the highest hill begins with a minimum size of 3 m stretched to a maximum size of 10 m. The grid-stretching ratio is 1.03. From the top of the highest hill to the up plane, the grid spacing is 10–60 m with the gird-stretching ratio of 1.04 [11].

All boundaries except wellhead are divided into five types as shown in Fig. 2, and the definitions are as follows: the inlet plane is set as a velocity-inlet boundary defined by vertical wind speed and temperature; the up plane is treated as a zero flux plane; the surface is defined as a no-slip wall, and wall function is also used to account for the inefficiency of grid resolution [26]; the others are outflow boundary.

An unsteady pressure-based solver is chosen to simulate an accidental scenario in two steps. First, simulation of the atmospheric flow is conducted without the wellhead jet; then, after a short time blowout, sour gas combustion and dispersion are calculated simultaneously. For every time step, the computing time requires about 5 min through parallel processing with Intel Quad-Core 2.83 GHz PC.

With tests of various time steps, the distribution of the contaminated areas will change slightly with the time step of less than or equal to 1.5 s.

SO$_2$ POISONING ASSESSMENT

Since that acute exposure to SO$_2$ is the main route of intoxication for the general public surrounding the well, the assessment is based on the dose–response relationship commonly used in acute toxicity research [21]. The toxic dose of potential exposure is evaluated by using the predicted concentration of SO$_2$ on the surface 146 cm above the ground. The height stands for the breathing zone, which is estimated by the average 136.7 cm shoulder height of standing Chinese adults [27] and a hemisphere of 15.24–22.86 cm diameter extending in front of the shoulder [28]. Considering that CFD results are discontinuous in time, linear interpolation of the concentration value is applied in integral intervals. The toxic dose is calculated as:

$$L = \int_0^{\tau_E} c^n d\tau = \sum_{k=0} \frac{1}{2}(c_k^n + c_{k+1}^n)\Delta\tau$$

(17)

Emergency response planning guideline (ERPG) levels developed by the American Industrial Hygiene Association are referenced as benchmarks for quantitative toxic environment assessment, which are based on acute toxicology data and designed to assist emergency response personnel planning for accidental chemical release to communities. For SO$_2$, the ERPG data are defined as exposure to airborne concentrations of 0.3, 3 and 15 parts per million (ppm) for up to 60 min [29]. For assessment facilitation, the equivalent ERPG concentration is used, which is associated with the estimated toxic dose through the following equation:

$$c_{eq} = (\frac{L}{60})^{1/n}$$

(18)

n is chosen as 1.0 according to the Center for Chemical Process Safety [21], and the contaminated area is graded into low (0.3–3 ppm), moderate (3–15 ppm), and high hazard (≥15 ppm) regions.

RESULT AND DISCUSSION

The basic information about the example well is 2.5 × 10^4 m^3/h of wellhead absolute open volume flux, 15.16% of volume fraction of hydrogen sulfide, and 5 min of ignition delay time.

Analysis of SO$_2$ Contaminated Areas

An illustration of the results with east winds is shown in Fig. 3. With high energy deriving from combustion and under the effects of wind, SO$_2$ disperses quickly and impacts a wide area that is irregular in shape. Within 15 min after ignition, the projection areas of high, moderate, and low hazard regions are 0.0051, 0.012, and 0.15 km^2, respectively. After 60 min, these regions increase to 0.028, 0.19, and 1.05 km^2 in area, respectively.

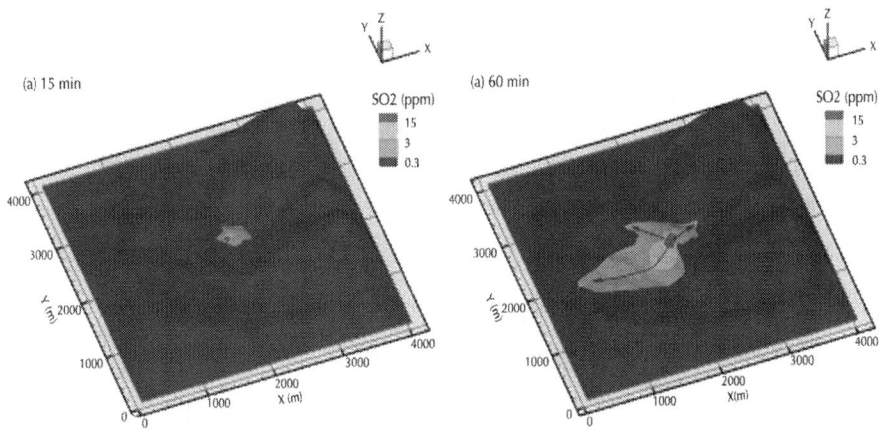

Figure 3: The distribution of SO$_2$ hazard regions at different time after ignition at 1.3 m/s east wind (a) 15 min and (b) 60 min.

In Fig. 3(a), the contaminated area is mainly distributed in the NE. A high hazard region surrounds the well with a maximal radius of 64 m. Moderate and low hazard regions head in a NE direction, with the farthest distances of 168 and 446 m, respectively. In Fig. 3(b), the contaminated area is mainly distributed in the SW and NW. High and moderate hazard regions extend to distances of 195 and 620 m in the SW, respectively. The maximal influence distances of low hazard regions are 1684 and 710 m in the SW and NW, respectively. In addition, the influence distances of high, moderate, and low hazard regions increase to 83, 207, and 535 m in the NE, respectively.

So the dispersion process could be estimated as: at an early stage, the plume accumulates in the valleys in the NE; then, the hazard regions spread downwind in SW and NW directions. The main trajectories of SO_2 include northeast valleys, and the northwest and southwest around the hills, which are indicated by arrows in Fig. 3(b).

The quantitative statistics of assessment results at different wind directions 60 min after ignition are shown in Table 1.

Table 1: Summaries of SO_2 contaminated areas

Wind direction	Contaminated area (km²)	Maximal influence distance (m)	Main trajectory
E	High: 0.028	High: SW 195 NE 83	(1) Northeast valleys
	Moderate: 0.19	Moderate: SW 620 NE 207	(2) Northwest
	Low: 1.05	Low: SW 1684 NW 710 NE 535	(3) Southwest
NE	High: 0.031	High: SW 184	(1) South valleys
	Moderate: 0.22	Moderate: S 569 SW 501	(2) Downwind areas
	Low: 2.02	Low: SW 1850 S 1,574	
N	High: 0.026	High: S 218	(1) Downwind areas
	Moderate: 0.17	Moderate: S 568	
	Low: 1.8	Low: S 1935	
NW	High: 0.072	High: SE 339	(1) South valleys
	Moderate: 0.36	Moderate: SE 784	(2) Downwind areas
	Low: 2.47	Low: SE 2023 S 1850	
W	High: 0.030	High: NE 190	(1) Northeast valleys
	Moderate: 0.13	Moderate: NE 420	(2) Downwind areas
	Low: 0.89	Low: E 1550 NE 807	
SW	High: 0.048	High: NE 318	(1) Downwind areas
	Moderate: 0.30	Moderate: NE 675	
	Low: 3.40	Low: NE 1995	
S	High: 0.036	High: N 230	(1) Northeast
	Moderate: 0.18	Moderate: NW 725 N 562	(2) Downwind areas
	Low: 2.49	Low: N 2033 NW 1897	
SE	High: 0.067	High: NW 340	(1) Downwind areas
	Moderate: 0.48	Moderate: NW 947	
	Low: 2.95	Low: NW 2362	

C	High: 0.013	High: R 80	(1) Vicinities of the well
	Moderate: 0.032	Moderate: N 264	(2) North valleys
	Low: 1.28	Low: N 1690	

Influence of Wind and Terrain

As the effects on SO$_2$ dispersion produced by wind are influenced by terrain and vice versa, conjoint analysis is carried out here.

In Fig. 4, the contaminated area of narrow ellipse is generated by east winds up to 5 m/s, quite different from Fig. 3(a). The hazard regions cover a much wider area under high wind speed. High, moderate, and low hazard regions are 0.0071, 0.013, and 0.35 km² in area, respectively, with corresponding maximal influence distances of 144, 294, and 1,705 m. The growth compared with the values in Fig. 3(a) is shown in Fig. 5. However, the impact is reduced. The average and maximal equivalent ERPG concentrations are 6.01 ppm versus 11.31 ppm and 239.43 ppm versus 339.81 ppm. In addition, these regions head straight west without obviously distributing in the upwind area. In this situation, the transportation of SO$_2$ by wind is primary, which drives the gas plume to cross the hills and suppress the gravity sedimentation of SO$_2$.

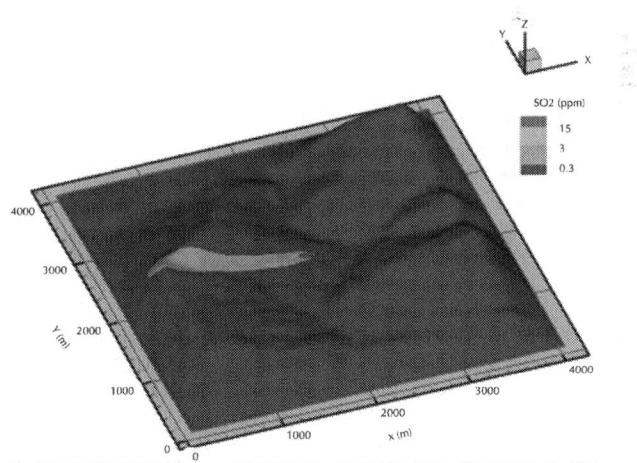

Figure 4: The distribution of SO$_2$ hazard regions at 5 m/s east wind after 15 min.

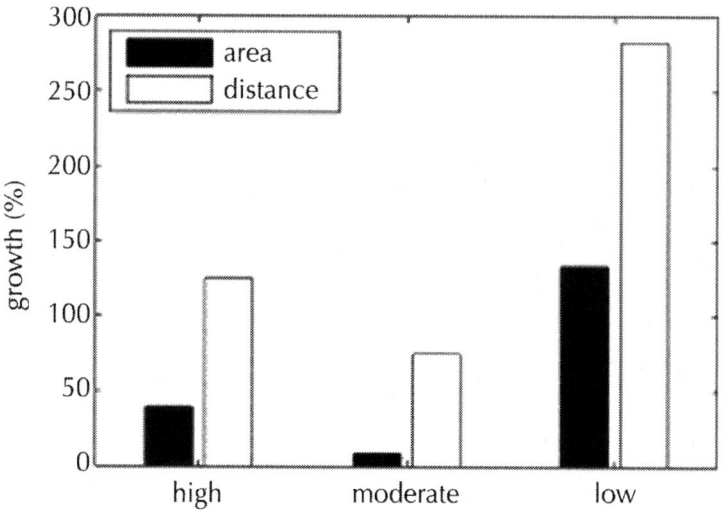

Figure 5: The growth of hazard area and distance at higher wind speed.

Comparing Fig. 6 to Fig. 3(b), significant differences exist in the distribution of the contaminated area, which are mainly caused by different terrain patterns. On the one hand, favorable terrain patterns such as valleys and basins, where wind speed is reduced and wind direction swirls, facilitate the sedimentation of SO_2. In Fig. 6, the winding valleys in the east are the largest part of the hazard regions. Though these areas are in the upwind direction in Fig. 3(b), the hazard exists there too, which is something to which attention should be paid. On the other hand, the plume is blocked and forced to change directions by the raised areas of the terrain. In Fig. 6, the downwind movement of SO_2 is restricted by a small hill in the east, which is in favor of the extension of hazard regions in the piedmont valleys. In Fig. 3(b), the plume direction turns to the southwest, where the terrains are relatively flat compared with those of the west.

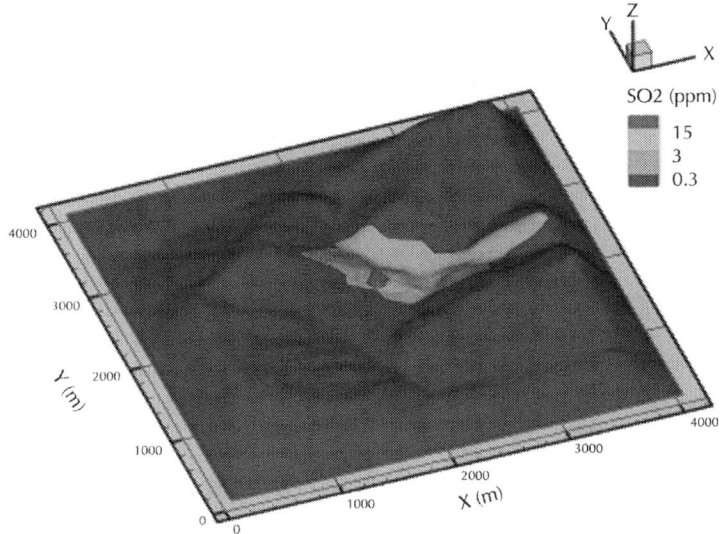

Figure 6: The distribution of SO₂ hazard regions at 1.3 m/s west wind after 60 min.

In summary, wind drives the hazard regions to spread downwind, which is the dominant factor for the distribution of hazard regions and could inhibit the effects of the terrains with the increase of wind speed. In addition, the shape of the hazard regions is changed by terrains under the effects of spatial aggregation and obstacles. However, the effects are hard to determine due to various characteristics of terrains and nonuniform distributions.

Distribution Features of the Contaminated Area

Under the interactions of wind and terrain, the hazard regions always spread in downwind areas, the front of the slopes, and depression areas such as valleys. However, without specific analysis, the appropriate estimation of the distribution is hard to make. In contrast to the situation in flat terrains, the hazard regions extend in several directions except for the downwind direction. The potential trajectories probably appear in the depression areas around the source and vicinities along the downwind direction where the terrains are flatter compared to the surrounding area.

Generally, worst-case dispersion conditions typically defined by low wind speed and stable atmospheric turbulence are chosen to do risk assessment. The hypothesis is that the affected regions predicted using this method could cover the largest potential accidental zone. However, the distribution of the contaminated area is so different from each other in complex terrains that a wide range of dispersion conditions should be included in addition to the worst-case scenario for practical applications such as emergency response planning.

Influence on Emergency Response

According to the analysis of distribution features of the contaminated area, terrain patterns conducive to gas dispersion and accumulation also facilitate human activities in the hills. Most people live in level ground suitable for building and agriculture. Roads and railways are always built through mountain passes, along riversides, and around mountainsides. These increase the possibility that the general public will be exposed to SO_2 in case of an accident.

During the emergency response, evacuation of the general public and transportation of huge amounts of emergency resources, such as equipment, and rescuers will be necessary. In view of the transportation system in the hills, the traffic load may increase to a level beyond the transport capacity. For example, there is only one road near the example well along the rivulet from west to north, with the capacity of one heavy vehicle. Considering that the rescue vehicles are mostly trucks, cranes, ambulances, and the like, and have to travel from the emergency response center located to the west of the well, the possibility of traffic congestion is quite high. As a result, the accident will be exacerbated by rescue delay and the evacuated are exposed to toxic gas for additional time due to a slow evacuation.

It can be concluded that the dispersed toxic gas is a great disadvantage to emergency response in relation to population distribution and transportation capacity in complex terrains and the bottleneck restriction comes from the transportation capacity.

Effectiveness in Controlling the Number of Deaths

Though the effectiveness in controlling the number of deaths by ignition of uncontrolled sour gas from well blowout is verified by the practical application [4], it is still valuable to provide another support for decision making in theory. Based on estimated toxic dose, probit for SO$_2$ lethal toxicity can be calculated according to [21]:

$$Y = -15.67 + \ln L \qquad (19)$$

In association with Eq. (18), the minimum lethal equivalent ERPG concentration is about 1.06×10^5 ppm. Calculation is done for all nine wind directions with combustion duration of 60 min. The results are all lower than that, meaning the theoretical probability of death is zero, which is quite in accordance with accident investigation.

CONCLUSIONS

To study the SO$_2$ poisoning due to ignition of uncontrolled sour gas flow of well blowout in hills, an integrated model is proposed to simulate the accident process of wellhead jet, sour gas combustion, and products dispersion, and to evaluate the consequences of SO$_2$ poisoning. The effective quantitative results provided by the simulations are useful for safety applications such as risk assessment and emergency response planning.

In complex terrains such as hills, the potential accidental consequences are varied with uncertain factors, mainly coming from the interactions of wind and terrain. As a result, the SO$_2$ will impact a wide area of an irregular shape in multiple directions during an accident. Considering that the real-time prediction of the toxic gas dispersion in complex terrains and atmospheric environment is still hard to obtain with current computer speeds, a well-prepared assessment of the potential hazard regions will be valuable to accident rescue. At least, preliminary accident analysis under the worst-case and most occurrences of dispersion conditions in eight wind directions should be performed.

In order to prevent the potential damage to the general public from the SO_2, a well-designed public protection plan should be added to the existing standards and guidelines in China, in which the related information is not so well specified. The adoptable measures include public notification, evacuation, and sheltering, which could be referred to Directive 071 issued by ERCB [3]. For the oil and gas companies and local governments, the emergency plan is deserved to pay attention to the traffic management to achieve a safe, timely, and orderly transportation of emergency resources and evacuation of the general public. To overcome the adverse effects of transportation capacity, it is suggested that temporary area monitors and portable shelters are provided along the main evacuation roads to help people take refuge and reduce the exposure to toxic gas.

It has been proven that SO_2 cannot cause death in practice and theory. So ignition decision should be made by on-site executive without hesitation under extreme conditions.

ACKNOWLEDGEMENTS

This research is supported by the National Key Science and Technology Programs of China for large oil and gas field exploitation (Project No: 2008ZX05017).

REFERENCES

1.   Technical Committee of Safety Standardization for Oil Industry, SY/T 5087-2005 Recommended practice for safe drilling operations involving hydrogen sulfide, National Development and Reform Commission, Beijing, China, 2005.

2.   SINOPEC Office of Science and Technology, Q/SH 0033-2007 Safety technique specification of gas well drilling and downhole operation in the Northeast of Sichuan, China Petrochemical Corporation, Beijing, China, 2007.

3.   ERCB, Directive 071 Emergency preparedness and response requirements for the petroleum industry, ERCB, Calgary, Canada, 2008.

4. Jianfeng Li, Bin Zhang, Yang Wang, Mao Liu, The unfolding of '12.23' Kaixian blowout accident in China, Saf. Sci. 47 (2009) 1107–1117.

5. Kaiji Zhou, Xin1 Liu, Zhaoxue Guo, Qiji Yuan, Lei Zhou, Study on the method predicting spurt speed of uncontrolled blowout at wellhead, Nat. Gas Ind. 26 (2006) 71–73.

6. Chen Jianguo, Pan Siming, Lu Yi, Yuan Hongyong, Numerical prediction and transportation study of poisonous gas transport on complex terrain, J. Tsinghua Univ. (Sci. Technol.) 47 (2007) 1–4.

7. Jianfeng Li, Bin Zhang, Mao Liu, Yang Wang, Numerical simulation of the largescale malignant environment pollution incident, Process Saf. Environ. (2009) 232–244, Prot. 87.

8. ERCB, ERCBH2S A Model for Calculating Emergency Response and Planning Zones for Sour Gas Facilities. Volume 1: Technical Reference Document, ERCB, Calgary, Canada, 2008.

9. P.A. Taylor, H.W. Teunissen, The Askervein hill project: overview and background data, Bound. -Layer Meteorol. 39 (1987) 15–39.

10. M. Btruer, N. Peller, Ch. Rapp, M. Manhart, Flow over periodic hills—numerical and experimental study in a wide range of Reynolds numbers, Comput. Fluids 38 (2009) 433–457.

11. A. Silva Lopes, J.M.L.M. Palma, F.A. Castro, Simulation of the Askervein flow. Part 2: Large-eddy simulations, Bound. -Layer Meteorol. 125 (2007) 85–108.

12. L.H. Hu, R. Huo, D. Yang, Large eddy simulation of fire-induced buoyancy driven plume dispersion in an urban street canyon under perpendicular wind flow, J. Hazard. Mater. 166 (2009) 394–406.

13. Paola Blotto, Michele Bonuccelli, Gianni Morale, Edoardo Dellarole, Mariano Falcitelli, Fabrizio Podenzani, Development of a integrated approach to the risk analysis of a blow-out accident, SPE 86704.

14. Lorenzo Borello, Michele Bonuccelli, Gianni Morale, The CFD approach for the risk analysis of a blowout event, SPE 108619.

15. FLUENT Inc. FLUENT 6.3 Help Document.

16. Kevin Mcgrattan, Bryan Klein, Simo Hostikka, Jason Floyd, Fire Dynamics Simulator (Version 5) User's Guide, NIST, Washington, DC, USA, 2008.

17. S.L. Brennan, D.V. Makarov, V. Molkov, LES of high pressure hydrogen jet fire, J. Loss Prev. Process Ind. 22 (2009) 353–359.

18. T. Allen, A.R. Brown, Large-eddy simulation of turbulent separated flow over rough hills, Bound. -Layer Meteorol. 102 (2002) 177–198.

19. Xian-Xiang Li, Chun-Ho Liu, Dennis Y.C. Leung, K.M. Lam, Recent progress in CFD modeling of wind field and pollutant transport in street canyons, Atmos. Environ. 40 (2006) 5640–5658.

20. A. Khelil, H. Naji, L. Loukarfi, G. Mompean, Prediction of a high swirled natural gas diffusion flume using a PDF model, Fuel 88 (2009) 374–381.

21. Center for Chemical Process Safety of the American Institute of Chemical Engineers, Guidelines for Consequence Analysis of Chemical Releases, American Institute of Chemical Engineers, New York, USA, 1999, pp. 216–261.

22. Sam Mannan, Lee's Loss Prevention in the Process Industries. Volume 1: Hazard Identification, Assessment and Control, 3rd ed., Elsevier ButterworthHeinemann Publications, Burlington, USA, 2005, pp. 16/220.

23. European Process Safety Centre, Atmospheric Dispersion, Institution of Chemical Engineers (IChemE), Rugby, UK, 1999, pp. 17–31.

24. K. Hanjalic, S. Kenjeres, Dynamic simulation of pollutant dispersion over complex urban terrains: a tool for sustainable development, control and management, Energy 30 (2005) 1481–1497.

25. F. Scargiali, E. Di Rienzo, M. Ciofalo, F. Grisafi, A. Brucato, Heavy gas dispersion modelling over a topographically complex mesoscale a CFD based approach, Process Saf. Environ. Prot. 83 (2005) 242–256.

26. Bert Blocken, Ted Stathopoulos, Jan Carmeliet, CFD simulation of the atmospheric boundary layer: wall function problems, Atmos. Environ. 41 (2007) 238–252.

27. China National Institute of Standardization and Information Classification and Coding, GB 10000-1988 Human dimensions of Chinese adults, National Bureau of Quality and Technical

Supervision, Beijing, China, 1988.

28.   Technical Committee of Safety Standardization for Oil Industry, SY/T 6137-2005 Recommended practices for oil and gas producing and gas processing plant operations involving hydrogen sulfide, National Development and Reform Commission, Beijing, China, 2005.

29.   Emergency Response Planning (ERP) and Workplace Environmental Exposure Level (WEEL) Committees, AIHA 2008 Emergency Response Planning Guidelines (ERPG) & Workplace Environmental Exposure Levels (WEEL) Handbook, American Industrial Hygiene Association, Fairfax, USA, 2008, pp. 1–27.

Chapter 3

Exploring the Chemical Aspects of Truck Tire Blowouts and Explosions

P.I. Dolez[a], C. Nohile[a], T. Ha Anh[a], T. Vu-Khanh[a],
R. Benoît[b], and O. Bellavigna-Ladoux[c]

[a]Ecole de technologie supérieure, 1100, rue Notre-Dame Ouest, Montréal, QC, Canada H3C 1K3
[b]Institut de recherche Robert-Sauvé en santé et en sécurité du travail, Montréal, QC, Canada
[c]Ecole Polytechnique, Montréal, QC, Canada

ABSTRACT

Truck tire blowouts and explosions, which account for a non-negligible number of occupational accidents with severe outcomes particularly in the transportation industry, are the subject of a literature review aimed at identifying the different processes involved, with a special focus on the chemical aspects of the phenomena. Tire blowouts and explosions associated with heat input are the result of the contribution of the thermal expansion of the air inside the tire, the thermal weakening of the tire structure, and three potential chemical reactions, namely pyrolysis, thermo-oxidation and combustion, leading to the degradation of the tire polymer matrix Tire pressure and air temperature data recorded during a torch-induced tire explosion experiment were analyzed and show a sharp transition at 97 °C between an initial pure thermal air expansion regime and a second regime, also linear and attributed to a combination of air expansion and reactions of thermo-chemical degradation of the tire materials. The findings also highlight the difficulties that are encountered when trying to devise preventive and corrective measures against tire blowouts and explosions.

INTRODUCTION

In Québec, between 1990 and 2004, two deaths every three years have been directly attributed to truck tire blowouts and explosions, as documented by the Commission de la santé et de la sécurité du travail (CSST, 2006) on their Web site. A much larger number of occupational accident files with a connection to tire blowouts and explosions can be located in the CSST central and regional data bank. In addition, some accounts can be found in the literature about such accidents, for example the company OTRACO providing a non-comprehensive list of 18 cases of earthmover tire explosions between 1981 and 2001 (OTRACO, 2004). Finally, personal communications with truck drivers and people involved in the transportation industry indicate that the phenomenon of tire blowout and explosion is far from rare, with potentially severe outcomes sometimes being barely avoided.

Tire blowouts can be associated with mechanical rupture of the tire, caused or not by overpressure, while explosions correspond to a much more violent process (Grogan, 1986). Indeed, for tire blowouts, the pressures involved are in the range of 0.7–1 MPa, whereas shock waves leading to explosions can create pressures exceeding 7 MPa (OTRACO, 2004). However, in both cases, the blast effect associated with energy release can severely hurt anyone located nearby. In addition, tire blowouts (which generally occur on moving vehicles) as contrasted to explosions (which most often take place on stationary vehicles) can create a loss of vehicle control leading to possibly severe consequences.

The two most frequent types of tire blowouts are bead rupture and zipper failure, both of which are generally associated with carcass damage due to under-inflated use (Grogan, 1986). For tire explosions, authors have identified a list of documented causes (Glenn, 1997 and Grogan, 1986): welding on rims or wheel assemblies, overheated brakes, electrical discharges from power lines and lightning strikes, fires, gross under-inflation of tires or overloading of vehicles, and the presence of combustible materials, contaminants, and absorbed fuel or inflammable fluids in the tire. All these situations can be associated with direct or indirect heat input into the tire.

In this article, an investigation of the phenomena occurring inside the tire because of abnormal heat input is carried out based on the literature resources and technical reports. Special focus is put on the possible chemical processes affecting the tire and created by the heat input. Then, the data obtained through a simulation of a real explosion case are analyzed. Finally, the chronology of the different events and their possible outcomes are proposed.

THE PROCESSES INVOLVED

During normal operating conditions, it has been measured that the temperature within the tire material sometimes exceeds 100 °C, in particular in the shoulder region (Schuring, 1980). When additional heat is brought to a tire, either directly through local heating by under-inflated use or a lightning strike, for example, or indirectly by way of a heated rim, the temperatures in both the tire material and in the air inside the tire are expected to increase. This temperature increase may generate three different processes in the tire: air expansion, loss of structural mechanical resistance, and chemical degradation.

Expansion of the Air inside the Tire

When air contained in an enclosed chamber such as a tire is subjected to a temperature increase, its volume increases. In the case of tires, such a volume increase is constrained by the low deformability of the rubber/reinforcement cable composite structure, leading to an increase in the pressure in the tire. In the ideal gas approximation and by neglecting the slight volume change allowed by the tire structure, the relationship between the air temperature T_i and pressure P_i at that temperature can be expressed as

$$\frac{P_i}{T_i} = \frac{P_0}{T_0},\text{ with the temperature in degrees Kelvin.}$$

For a truck tire inflated to 0.62 MPa at 20 °C, which is the pressure recommended by manufacturers, the pressure will reach 0.79 MPa at 100 °C and 1.2 MPa at 300 °C. For a new tire, since a safety factor of 3–8 is usually included by manufacturers (Frates, 2000), the risk of tire blowout is limited under the sole effect of thermal expansion of the air. However, if the tire has a structural defect or has suffered prior damage that have weakened its mechanical resistance, the maximum pressure it can sustain without blowing out can be significantly reduced.

Thermal Weakening of the Tire's Mechanical Structure

Since a tire is a composite structure, the change in its mechanical resistance due to an increase in temperature may arise both from changes affecting the individual components themselves or the links between them. However, in the first place, rims, and especially aluminum alloy rims, even if they are not considered as a part of the tire, may also be sensitive to temperature: for example, above 315 °C, the physical properties of some aluminum rims are strongly affected (Dupras and L'Épicier, 2002), making them susceptible to permanent deformation under normal load. With such a deformed rim, chances of tire/rim separation are high, with the associated blowout-type consequences.

Temperature changes also physically affect the rubber constituting the tire matrix. For example, between 25 °C and 100 °C, a temperature range well within the normal operating conditions of tires, it has been shown that rubber undergoes a slight thermal softening, translating into a decrease in the storage modulus measured by dynamical mechanical analysis (Burlett, 2004). Between 95 °C and 125 °C, microstructural changes within vulcanized rubber may create substantial stress relaxation, material softening, permanent set, and creep (Wineman et al., 2003), which affect the rubber's mechanical behavior.

To improve adherence between the reinforcement cables and the rubber matrix, steel cables are coated with bronze or brass, and adhesion is achieved through non-stoichiometric metallic sulfides created by the reaction of the active sulfur in the rubber with the oxide layer on the coated steel (Su and Shemenski, 2000). This bonding layer appears to be very sensitive to temperature, as revealed by the decrease in pull-out force observed when the coated steel cable/rubber composite is subjected to overcuring at a temperature of 141 °C (Chandra et al., 1996).

Finally, studies on the effect of temperature on whole tire composite behavior have shown a large decrease in the slope of tensile stress–strain curves between 40 °C and 160 °C (Tan et al., 2003). Even if the loss in tire mechanical resistance due to thermal weakening of the structure is by itself insufficient to lead to tire rupture, these damaging effects may combine with other processes to create the conditions corresponding, for example, to tire blowout. In addition, the thermal weakening of the tire structure may have a synergetic effect on the normal aging process due to environmental and service conditions: for example, it is well known that oxidation becomes an autocatalytic reaction at elevated temperature, with oxygen absorption deviating from its initial linear variation with time (Hofmann, 1989).

Thermo-chemical Degradation of Rubber

The polymer matrix of truck tires is typically composed of 56% rubber, 26% carbon black particles, and 18% various additives (Ciullo and Hewitt, 1990). The rubber is about 80% natural rubbers, with the rest corresponding to various synthetic copolymers of styrene and/or butadiene. The total tire weight also includes about 17% textile reinforcement and steel cables (Gabor et al., 2001).

When the polymer matrix is subjected to high temperatures, various chemical reactions may occur, leading to the thermo-chemical degradation of the material. Most documents circulating in the truck transportation industry refer to pyrolysis as the culprit chemical reaction (Anon, 2000, Breault, 2005 and Glenn, 1997). Pyrolysis takes place in the absence of oxygen under the sole action of heat. Thermogravimetric analyses of truck tire materials have measured initial degradation temperatures by pyrolysis as low as 185 °C (Chen et al., 2001). However, since most truck tires are inflated with air, which contains a large percentage of oxygen, thermo-oxidation reactions may also occur on the inner surface of the tire. It has been shown that the presence of oxygen leads to a decrease in the degradation temperature as compared to inert gas (Verdu, 1984). Finally, combustion may take place with enough oxygen for complete oxidation of the chemical species.

A more detailed presentation of the particularities of these three chemical processes will be provided in the following section of this article. However, one of the common characteristics of all three reactions is that they produce flammable gases. In the first place, when the thermo-chemical reactions take place on the inner surface of the tire, the gases they produce add to the increase in pressure inside the tire due to thermal expansion of the air. It has been calculated that the chemical degradation of only 20 g of rubber produces enough gas to reach the standard blowout pressure of tires (Thomassin and Blais, 2004). In addition, with an auto-ignition temperature situated around 430 °C; these flammable gases can generate an explosion at high temperature if their concentration inside the tire is above 1–8% while the oxygen concentration is above 5.5% (Ontario Natural Resources Safety Association, 1996).

Among the other products of the chemical degradation of tires, sulfur and carbon black particles may also produce explosions, with auto-ignition temperatures of 200 °C for carbon (Thomassin and Blais, 2004) and 190 °C for sulfur (Le ministre de l'environnement, 1996). For sulfur, which is present at 1–2% in tires, the critical concentration is equal to 30 mg/l (Le ministre de l'environnement, 1996).

It is also worth mentioning that the presence of combustible or flammable contaminants in the tire has been put forward as the initiator of or contributor to tire blowouts and explosions (Glenn, 1997). While the most common contaminants are petroleum-, solvent- and silicone-based lubricants as well as gasoline and ether used by some tire fitters as the sealant, one case has been documented in which the diffusion of methanol from the packing wood left inside a haultruck tire was the cause of an explosion (OTRACO, 2004).

THE CHEMICAL REACTIONS

In this section, more detailed information is provided on the three types of thermo-chemical reactions that have been identified as potential contributors to the final blowout or explosion event.

Pyrolysis

Under the sole effect of heat, the pyrolytic chemical reaction induces the uncatalyzed cleavage of covalent bonds, most often hydrogen bonds, leading to the creation of free-radicals (Verdu, 1984). Degradation proceeds through chain propagation. Such a reaction occurs either in the absence of oxygen, or when the oxygen input is diffusion-limited. For sulfur-crosslinked elastomers, which make up most of the tire polymer matrix, pyrolysis has been shown to lead to an increase in the crosslink density between 100 °C and 180 °C, followed by a destruction of the sulfur crosslinks between 180 °C and 220 °C (Burlett, 2004).

Thermogravimetric analysis of the pyrolysis degradation of tires has revealed three distinct reactions (Aylon et al., 2005 and Leung and Wang, 1998): between 185 °C and 310 °C, a large peak was attributed to the degradation of organic additives; then, a narrow peak at 380 °C was associated with natural rubber degradation; finally, the last peak between 350 °C and 490 °C was attributed to the synthetic elastomers present in the tire, for example butadiene (465 °C) or styrene–butadiene (444 °C) rubber.

Three types of products are obtained from the degradation of tires (Clark et al., 1993 and Cunliffe and Williams, 1998): solid matter (35–40% excluding textile and cable reinforcement material), oils (38–55%), and gas (10–30%). Solid residues include carbon black, ash, and inorganic matter such as zinc oxide, carbonates and silicates, with a gross calorific value situated around 30 MJ/kg (Gonzalez et al., 2001). Oil products consist of 6- to 24-carbon organic compounds, comprising several aliphatic and aromatic hydrocarbons (Laresgoiti et al., 2004). Their gross calorific value of 40–42 MJ/kg is even higher than that specified for commercial heating oils. The gases produced by pyrolytic degradation of tire material consist mostly of paraffins and olefins with carbon numbers ranging from one to five (Clark et al., 1993). The gross calorific value of pyrolytic gas is situated between 35 and 40 MJ/m^3, similar to that of natural gas. The products of tire pyrolysis thus display the characteristics of efficient combustible materials.

Thermo-oxidation

Thermo-oxidation takes place when oxygen is present but not in sufficient quantity for complete combustion. However, even in the presence of oxygen, both pyrolysis and thermo-oxidation may coexist since non-oxidative reactions such as pyrolysis are very fast, and oxidative processes are controlled by oxygen availability (Verdu, 1984). Compared to pyrolysis, thermo-oxidation is also a free-radical process, but it may occur at a lower temperature and be much more exothermic. In addition, at sufficiently high temperature, oxygen absorption becomes autocatalytic, with one free-radical generating three free-radicals (Hofmann, 1989), leading to reaction acceleration.

The thermo-oxidative degradation of styrene–butadiene rubber, the most widely used synthetic rubber in tires, has been studied by thermogravimetry at various oxygen concentrations (Chen et al., 1997) As the oxygen/nitrogen ratio is increased from 0% to 20%, the main degradation peak at 485 °C narrows, indicative of shorter reaction times, and shifts slightly towards lower temperatures. In addition, a second peak appears at 527 °C. This second peak was attributed to the oxidation of the carbonaceous residues produced by the first stage of the reaction (Conesa et al., 1998).

As for pyrolysis, the thermo-oxidation products consist of solids, oils and gases, with similar highly combustible properties. However, the presence of oxygen leads to an increase in the fraction of gas produced vs. oils and solids (Lee et al., 1995). In addition, some gases are found only in the case of thermo-oxidation, such as carbon monoxide, carbon dioxide, hydrogen and sulfur dioxide (Clark et al., 1993). Finally, the thermo-oxidation signature can also be identified in the oil and solid products, with the presence of hydroxyl and carbonyl groups originating from the decomposition of peroxy radicals (Burlett, 1999).

Combustion

The combustion process takes place when enough oxygen is available to fully oxidize the species. Tire combustion proceeds through four stages. Between 200 °C and 480 °C, a first decomposition can be observed, similar to pyrolysis but with slightly different kinetic parameters and a lower reaction temperature (Conesa et al., 1998). Next, the carbonaceous residues of the first reaction are oxidized between 480 °C and 500 °C. During the third step between 600 °C and 650 °C, the inorganic fraction of the tire material, such as calcium carbonate and zinc oxide, are decomposed. Finally, combustion of solid carbon takes place above 800 °C (Choi et al., 2001).

Unlike pyrolysis and thermo-oxidation, combustion residues include only solids and gases (Kim et al., 1994). In the case of complete combustion, gas products are composed of carbon dioxide, sulfur dioxide, nitrogen and water (Conesa et al., 1998). Carbon monoxide and nitrogen monoxide may also be present if the combustion is incomplete.

POSSIBLE SCENARIO FOR AN EXPLOSION

A scenario has been proposed in the literature that illustrates the development of the conditions leading to the explosion of a truck tire (see Fig. 1) (Ontario Natural Resources Safety Association, 1996). When the truck moves, the temperature of the tire material (black curve) first increases (marker 1), then equilibrates (marker 2) to its normal operating value, which depends among other things on tire composition, truck speed, ambient temperature, truck load and tire pressure, and varies for the different tire parts and throughout the thickness of the tire. Once a heat source is activated near the tire (marker 3) and after a certain delay due in part to the low thermal conductivity of rubber, the temperature starts to increase. When the temperature reaches the tire material's degradation temperature (marker 4), the production of flammable gases begins (gray curve). As the temperature further increases, it eventually reaches the auto-ignition temperature of the flammable gases (or of any other flammable material produced by the chemical degradation reaction inside the tire). If the concentration of theses flammable species has not yet reached the critical value, no explosion occurs.

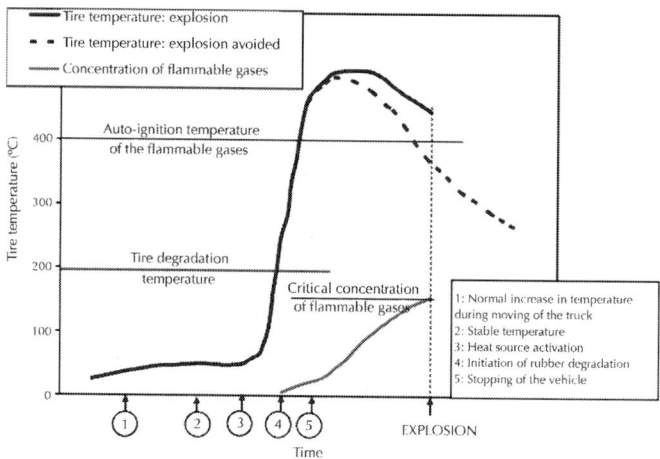

Figure 1: Evolution in the phenomena on a time scale (from (Ontario Natural Resources Safety Association, 1996)).

In the case illustrated in Fig. 1, the truck comes to a stop at a time labeled with marker 5. Let us imagine that it corresponds to a removal of the abnormal heat source. However, this does not stop the temperature from increasing. In fact, the temperature increase may even be expected to accelerate since the tire is no longer being cooled down by the air circulation induced by truck movement (Pyrotech BEI, 2002). In addition, tire degradation chemical reactions are exothermic and, after a certain time, can keep themselves going without any additional heat input. When the tire temperature finally starts to decrease, one of the two scenarios illustrated on the graph in Fig. 1 can occur. In the first case (dotted black curve), the temperature inside the tire drops below the auto-ignition temperature of the flammable gases before their critical concentration has been reached In that case, the explosion is avoided. According to the second scenario (solid black curve), the temperature inside the tire is still above the auto-ignition temperature of the flammable gases when their critical concentration is reached. If more than 5.5% oxygen is present inside the tire, an explosion occurs.

It is clear that tire blowout may have occurred at any moment during the course of the events prior to the explosion if the pressure inside the tire had overcome its mechanical resistance. In addition, other scenarios are possible, for example the explosion occurring before the temperature had any chance to decrease. It is also worthwhile noting that, in the case of the avoided explosion scenario, the gases produced by the chemical reaction will remain in the tire if not purged: in the next instance of abnormal heating, the level of flammable gases needed to reach the critical concentration will be reduced accordingly, and therefore increase the risk of explosion.

ANALYSIS OF A REAL CASE

An experiment of controlled explosion of a truck tire (size 11R22.5) due to a 35-s heating of the rim with a torch (see Fig. 2) was performed by Michelin and recorded on a movie (Welding on Wheels.wmv, Paul Labadie (Desharnais Centre du camion, Québec), François Beauchamp (Michelin, Canada), private communication, October 2005). The pressure and air temperature inside the tire were recorded during the experiment and the corresponding displays are also shown in the movie. Fig. 3 presents the variation of these two parameters as a function of the time elapsed since the beginning of torch heating of the rim. The pressure values reached the limit of the gauge display before the explosion occurred, which explains why the last two pressure data points are missing in Fig. 3. Both the pressure and air temperature inside the tire remained relatively stable for the first 110 s. At that point in time, which corresponds to 75 s after the torch was turned off, both pressure and temperature started increasing more and more rapidly. The explosion occurred at 132 s from the beginning of the experiment, i.e., a little over 1.5 min after the torch had been turned off.

Figure 2: Extract from the movie showing the experiment set-up with the torch heating the tire rim and the temperature and pressure displays (Weld-

ing on Wheels.wmv, Paul Labadie (Desharnais Centre du camion, Québec), François Beauchamp (Michelin, Canada), private communication, October 2005).

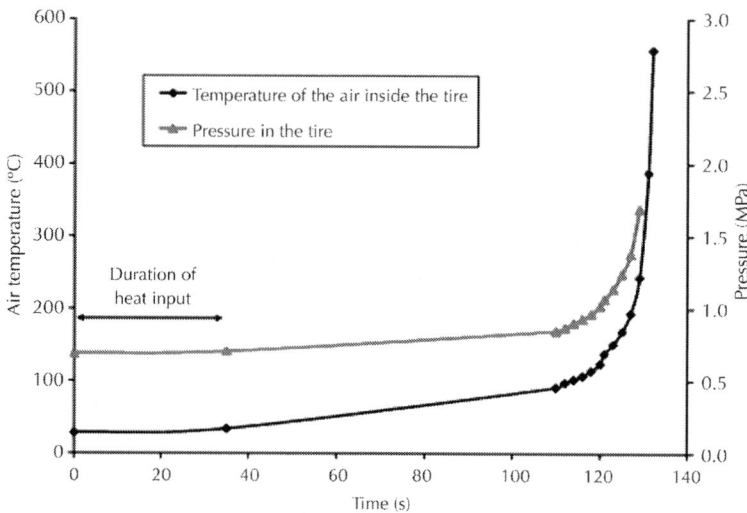

Figure 3: Variation in the measured temperature and pressure as a function of the elapsed time.

Fig. 4 shows the variation in the recorded pressure as a function of the temperature inside the tire during the experiment. A linear relationship attributed to thermal expansion of the air is observed up to a temperature of 97 °C. Above that point, the pressure/temperature variation also follows a linear behavior but with a much greater slope. The change in slope is attributed to the initiation of a new process, a thermo-chemical reaction, which adds to the thermal expansion of the air and eventually leads to the ultimate explosion of the tire. It must be mentioned that, since the pressure gauge limit was exceeded some time before the explosion occurred, it has not been verified that this pressure/temperature linear variation was sustained up to the explosion. However, such a linear variation in the quantity of gas produced as a function of the reaction temperature has been observed both for the pyrolysis and thermo-oxidation degradation of tires (Gonzalez et al., 2001 and Lee et al., 1995).

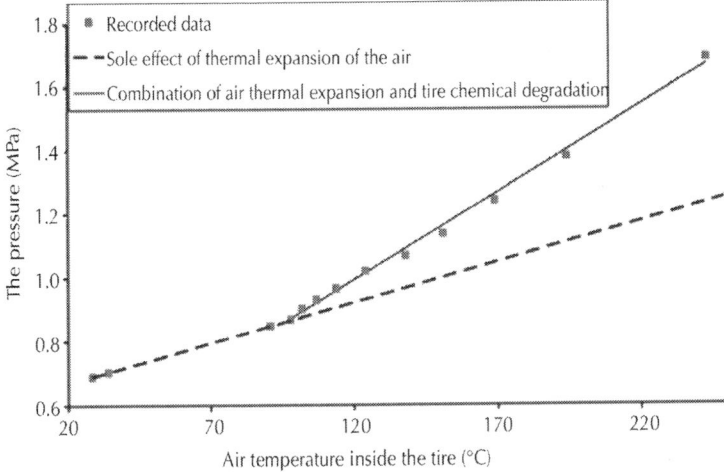

Figure 4: Variation in the recorded tire pressure as a function of the temperature of the enclosed air.

This controlled explosion experiment shows that the shift from pure thermal expansion to the appearance of a contribution from a thermo-chemical degradation reaction can be clearly identified by a sharp change in the slope of the pressure/temperature linear variation. However, in the case of that experiment, the time between this turning point and the explosion was only 20 s, which is rather short for performing much avoidance action. Another important piece of information that emerged from this experiment concerns the air temperature inside the tire at which the thermo-chemical reaction started: the recorded 97 °C value is relatively close to the air temperatures reported for normal service conditions, for example 88 °C at 50 mph for a truck tire (Richey et al., 1956).

It must be mentioned that a large range of times between the event generating the additional heat input and the resulting tire blowout or explosion can be identified in the literature and in accident reports as illustrated below. At one extreme on the scale, in this torch heating experiment, 1 min and 37 s elapsed between the torch being turned off and the explosion. With lightning strikes, the enormous amount of energy involved usually leads to instantaneous explosions (OTRACO, 2004). On the other side, in the case of an accident caused by a leak in a truck's braking system, which led to the explosion of a tire and the death of the mechanic trying to repair it (Thomassin and Blais,

2004), it was estimated that the explosion occurred more than 1 h after the vehicle came to a full stop. Finally, various organizations, for example the Natural Resources, Mines and Energy Department of the Queensland government in Australia and the Quebec Fire Safety Service in Canada, recommend an isolation period of 24 h when a risk of tire explosion has been identified (Breault, 2005 and Minahan, 2004).

DEVELOPMENT OF THE PROCESSES

An alternative representation to the developing processes illustrated in Fig. 1 is proposed in Fig. 5: the diagram follows the progression of the phenomena by means of the evolution in tire temperature and pressure, and focuses on the different processes leading either to a blowout or an explosion.

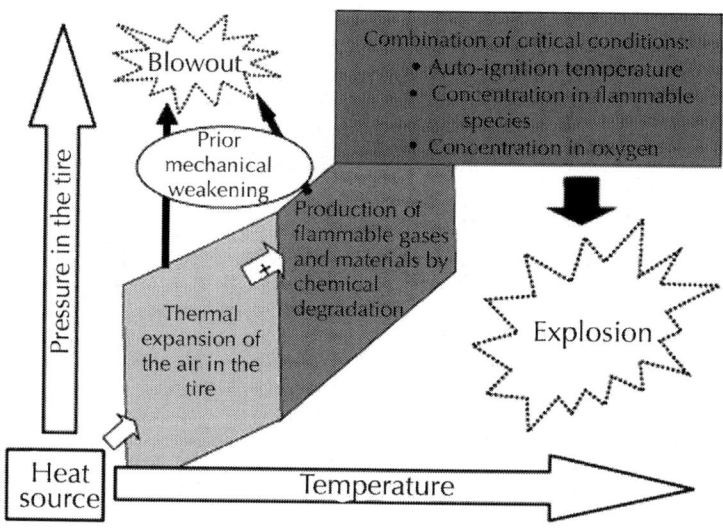

Figure 5: Evolution in the processes leading to tire blowout or explosion.

When the tire temperature increases past its normal operating conditions because of an abnormal heat source, the first process that occurs is the thermal expansion of the air inside the tire. This creates an increase in tire pressure, which can lead to a blowout, especially if the tire has already suffered damage or has a structural defect. As a second step, if the temperature reaches a certain level, while thermal expansion of the air continues, degradation reactions of the rubber in the tire begin, generating flammable gases and materials inside the tire. These gases add to the quantity of air initially in the tire and contribute to the pressure rise inside the tire. As a consequence, the risk of a blowout increases, and increases further as chemical degradation proceeds, with both the produced gases and air undergoing thermal expansion. The third and ultimate stage occurs when three conditions are met: the auto-ignition temperature of any of the flammable species in the tire is reached as well as the critical concentration of that flammable species while the concentration of oxygen is above 5.5%. At that point, an explosion takes place, generating shock waves and pressures of about 7 MPa.

CONCLUSIONS

A literature survey has been carried out to profile the state of current knowledge on the phenomena of tire blowouts and explosions, in particular on the chemical processes involved. Since a heat source produces a temperature increase in both the air inside the tire and in the tire material itself, three processes have been identified as possible contributors to tire rupture: thermal expansion of the air inside the tire, thermal weakening of the tire structure, and chemical degradation of the tire polymer matrix. Three chemical reactions may be responsible for tire degradation, namely pyrolysis, thermo-oxidation, and combustion, depending on the amount of oxygen available. Studies on truck tire pyrolysis have recorded degradation temperatures as low as 185 °C, while it has been shown that the presence of oxygen in thermo-oxidation and combustion reactions should translate into a reduction in that temperature. This degradation-onset temperature is relatively close to the normal service temperatures of some parts of the tire, which have been shown to reach and exceed 100 °C.

In the chronology of the events leading to tire blowout or explosion caused by abnormal heat input, the first step as the temperature increases corresponds to the thermal expansion of the air inside the tire, which can lead to a blowout if the structure of the tire is already weakened. Then, when the tire temperature reaches the initiation of the degradation reactions, the production of flammable species starts, including gases, which further increase the risk of a blowout. The final stage takes place when the three critical conditions for an explosion are met, i.e., the auto-ignition temperature and critical concentration of a flammable species, and an oxygen concentration above 5.5%. Such an explosion can occur even after the heat source has been removed, since the chemical reactions involved are highly exothermic. The stopping of the vehicle further increases the risk of explosion since the heated tire is no longer being cooled down by air circulation.

An analysis of pressure/temperature data recorded during the simulation of a tire explosion caused by torch heating of the rim has shown a sharp change in regime occurring at an air temperature inside the tire of 97 °C. Below that temperature, the linear temperature/ pressure variation associated with thermal air expansion was observed. Above it, a linear behavior was still recorded but with a much larger slope, indicating that a gas-generating chemical reaction was contributing to the pressure increase. The air temperature inside the tire at which chemical reaction initiation is observed is relatively close to the upper limit of air tire temperature during normal operating conditions. In addition, the time between the detected initiation of the chemical reaction and the explosion is only 20 s in that particular case of torch heating of the rim, which are rather short for corrective or reactive actions. However, in other documented cases, much longer times have been reported between the removal of the heat source and the occurrence of the explosion, for example more than 1 h in the case of a malfunctioning braking system.

This paper has identified the processes involved in tire blowout and explosion phenomena. It has also pinpointed some of the obstacles that will have to be considered when dealing with preventive and corrective measures for tire blowouts and explosions.

ACKNOWLEDGEMENTS

This work was funded by the Institut de recherche Robert-Sauvé en santé et en sécurité du travail. The authors also want to thank Paul Labadie of Desharnais Centre du camion and François Beauchamp of Michelin – Canada for providing them with the movie of the explosion experiment.

REFERENCES

1.   Anon, 2000 Tyre explosions can kill Minesafe 11 (3), 2.

2.   Aylon, E., Callen, M.S., Lopez, J.M., Mastral, A.M., Murillo, R., Navarro, M.V., 2005 Assessment of tire devolatilization kinetics Journal of Analytical and Applied Pyrolysis 74 (1–2), 259–264.

3.   Breault, M., 2005 UN risque a` connaitre Services de se´curite´ incendie au Que´bec <http://www.securiteincendie.com/> (retrieved 31.10.2005).

4.   Burlett, D.J., 1999 Studies of elastomer oxidation via thermal analysis. Rubber Chemistry and Technology 72 (1), 165–173.

5.   Burlett, D.J., 2004 Thermal techniques to study complex elastomer/filler systems. Journal of Thermal Analysis and Calorimetry 75 (2), 531–544.

6.   Chandra, A.K., Mukhopadhyay, R., Bhowmick, A.K., 1996 Studies of dynamic adhesion between steel cord and rubber using a new testing method Polymer Testing 15 (1), 13–34.

7.   Chen, K.S., Yeh, R.Z., Chang, Y.R., 1997 Kinetics of thermal decomposition of styrene butadiene rubber at low heating rates in nitrogen and oxygen Combustion and Flame 108, 408–418.

8.   Chen, J.H., Chen, K.S., Tong, L.Y., 2001 on the pyrolysis kinetics of scrap automotive tires Journal of Hazardous Materials 84 (1), 43– 55.

9.   Choi, B.-C., Choi, W., Lim, M.-T., 2001 Burning used automotive tires for activated carbon JSME International Journal, Series B: Fluids and Thermal Engineering 44 (1), 133–139.

10. Ciullo, P.A., Hewitt, N., 1990. Rubber In: Library, P.D. (Ed.), The Rubber Formulary. Noyes Publications/William Andrew Publishing, New York, pp. 1–72.

11. Clark, C., Meardon, K., Russel, D., 1993. Scrap tire pyrolysis. In: Corporation, N.D. (Ed.), Scrap Tire Technology and Markets. Commission de la sante´ et de la se´curite´ du travail (CSST), 2006   <http://www.csst.qc.ca/portail/fr/actualites/rap_enquete/rap_enquete.htm>.

12. Conesa, J.A., Font, R., Fullana, A., Caballero, J.A., 1998. Kinetic model for the combustion of tyre wastes. Fuel 77 (13), 1469–1475.

13. Cunliffe, A.M., Williams, P.T., 1998. Composition of oils derived from the batch pyrolysis of tyres. Journal of Analytical and Applied Pyrolysis 44 (2), 131–152.

14. Dupras, A., L'E´ picier, A., 2002 Accident investigation report: Rapport d'enque^te d'accident: Accident mortel survenu a` un travailleur le 3 Juillet 2001 au centre marai^cher Euge`ne Guinois Jr Inc. soit au 555, 4e`me rang a` Sainte-Clothilde-de – Cha^teauguay. Commission de la sante´ et de la se´curite´ du travail du Que´bec.

15. Frates, W.S., 2000. Tire inflation accidents. <http://www.tireaccidents.com/tire_inflation_accidents.htm>.

16. Gabor, J.L., Wall, J.F., Rodgers, M.B., 2001. Overview of medium radial truck and off the road tire technology. In: Proceedings of the 2001 American Chemical Society Spring meeting, Rubber division, Providence, Rhode Island.

17. Glenn, W., 1997 Exploding tires the hazard nobody's heard of. Occupational Health and Safety Canada Magazine 13 (2), 42–48.

18. Gonzalez, J.F., Encinar, J.M., Canito, J.L., Rodriguez, J.J., 2001 Pyrolysis of automobile tyre waste – Influence of operating variables and kinetics study. Journal of Analytical and Applied Pyrolysis 58, 667–683.

19. Grogan, R.J., 1986 an investigator's guide to tire failure North Florida University, Jacksonville (128p.).

20. Hofmann, W., 1989 Aging and aging protectors In: Hofmann, W. (Ed.), Rubber Technology Handbook. Oxford University Press, pp. 264–268.

21. Kim, J.R., Lee, J.S., Kim, S.D., 1994 Combustion characteristics of shredded waste tires in a fluidized bed combustor Energy 19 (8), 845– 854.

22. Laresgoiti, M.F., Caballero, B.M., de Marco, I., Torres, A., Cabrero, M.A., Chomon, M.J., 2004. Characterization of the liquid products obtained in tyre pyrolysis. Journal of Analytical and Applied Pyrolysis 71 (2), 917–934.

23. Lee, J.M., Lee, J.S., Kim, J.R., Kim, S.D., 1995 Pyrolysis of waste tires with partial oxidation in a fluidized-bed reactor. Energy 20 (10), 969–976.

24. Le ministre de l'environnement, 1996 Circulaire DPPR/SEI du 26 juillet 1996 relative a` la nomenclature des installations classe´es pour la protection de l'environnement Class ement du soufre, from <http://aida.ineris.fr/textes/circulaires/text0145.htm>.

25. Leung, D.Y.C., Wang, C.L., 1998 Kinetic study of scrap tyre pyrolysis and combustion. Journal of Analytical and Applied Pyrolysis 45 (2), 153–169.

26. Minahan, P., 2004 Tyre Fire, Pyrolysis and Explosions Safety Bulletin 47 Queensland Government, Natural Resources, Mines and Engineering.

27. Ontario Natural Resources Safety Association (ONRSA), 1996 Technical report: Tire explosions due to pyrolysis.

28. OTRACO, 2004 Final report to Hamersley Iron on catastrophic tyre incident on haul truck 06H1 at Channar on April 10, 2004.

29. Pyrotech BEI, 2002 Rapport d'expertise Pyrotech BEI (No. 003364).

30. Richey, G.G., Hobbs, R.H., Stiehler, R.D., 1956. Temperature studies of air in truck tire. Rubber Age 79 (2), 273–276.

31. Schuring, D.J., 1980 Rolling loss of pneumatic tires. Rubber Chemistry and Technology 53 (3), 600–727.

32. Su, Y.-Y., Shemenski, R.M., and 2000 Impact of oxide characteristics on coatings of steel reinforcement to rubber adhesion In: Wire Association International Inc. (Eds.), Conference Proceedings Wire and Cable Technical Symposium (WCTS). Nashville, TN, USA, pp. 50–59.

33. Tan, H.-F., Du, X.-W., Wei, Y.-T., Yum, Y.-J., 2003 Mechanical properties of cord-rubber composites and tire finite element analysis. Vehicle System Dynamics 40 (Suppl.), 161–174.

34. Thomassin, M., Blais, Y., 2004 Accident investigation report: Accident mortel survenu a` un travailleur le 2 avril 2004 dans la re´serve faunique des Laurentides.

35. Verdu, J., 1984. Vieillissement des plastiques: Paris: Association francaise de normalisation, Diffuse´ par Eyrolles.

36. Wineman, A., Jones, A., Shaw, J., 2003 Thermomechanics of elastomers undergoing scission and crosslinking at high temperatures Tire Science and Technology 31 (2), 68–86.

Intensity and Efficiency of Spray Fuel-Fed Well-Mixed Adiabatic Combustors

Daniel E. Rosner[a], Manuel Arias-Zugasti[a, b], and Michael Labowsky[a]

[a]Department of Chemical Engineering, High Temperature Chemical Reaction Engineering Laboratory and Yale Center for Combustion Studies, Yale University, New Haven, CT06520-8286, USA

[b]Departamento de Física Matemática y de Fluidos, Facultad de Ciencias UNED, Apdo: 60141, 28080 Madrid, Spain

ABSTRACT

Motivated by the insights it can provide, we revisit the classical problem of liquid fuel-fed idealized steady-flow combustors. New quadrature-based results are presented for the theoretical combustion *intensity* and corresponding *efficiency* for well-stirred adiabatic vessels fed with a prescribed polydispersed spray. Each droplet of the spray is assumed to evaporate according to a non-quasi-steady (non-QS) gas-phase energy/ mass diffusion-controlled rate for the pseudo-single-component fuel. As a byproduct, we calculate the complete droplet size distribution (DSD) function exiting the chamber, of interest for the design of downstream components. We explicitly assume that the volumetric rate of chemical energy release in such "primary" combustion chambers is controlled by the liquid fuel physical vaporization process (with negligible lags due to propellant droplet heat-up or vapor-phase ignition). In this instructive asymptotic limit, two decisive non-dimensional parameters are shown to be: (1) a vaporization Damköhler number (defined by the ratio of the mean residence time of the chemically reacting vapor mixture in the combustion space, to the reference value of the vaporization lifetime of a droplet with the injector-Sauter-mean diameter, and (2) a single dimensionless non-QS parameter. Illustrative numerical results for a kerosene-like fuel burning in air at pressures up to 24 atm are displayed for the case of a log-normal feedstream DSD with a range of spreads. Our results reveal the existence of an optimum vaporization Damköhler number which maximizes the combustion intensity—with maximum intensities, occurring well before nearly complete fuel evaporation, being quite sensitive to the non-QS parameter at high pressures. These deliberately idealized mathematical model results, spanning more than a 1000 combinations of operational parameters, set instructive bounds to the achievable performance of "real" spray combustors. Even without tractable enhancements (see Section 5.2), this approach can be used to economically map the sensitivity of spray combustor performance to a large number of important design and control parameters.

INTRODUCTION AND OBJECTIVES

Whether intended for a combustion turbine, a liquid propellant rocket motor, or as the first step in an industrial-scale chemical synthesis process (e.g., H_2SO_4, H_3PO_4, see e.g., Rosner, 1997), spray fuel-fed combustors share certain common characteristics. It is therefore instructive to develop simple mathematical models, of course having a rational *physical* basis, to describe their performance. Useful engineering models will provide potentially important information about how combustor performance will depend on (at least some) such factors as: geometry (chamber volume, shape), inlet conditions (liquid phase and gas phase), liquid fuel properties, and injector ("atomizer") performance. This information can, of course, be used for purposes of engineering design, predicting the consequences of, say, fuel substitution, and even certain types of model-based *control* of an existing design.

While the approach of full three-dimensional multi-phase computer modeling is evidently in widespread simulation and/or design use today (see, e.g., Menon and Patel, 2006, Wittig et al., 2002 and Tolpadi et al., 2000, having by now benefited from significant cumulative investments over more than a 40-year period), in fact a broad spectrum of "hybrid" approaches, including so-called "modular" models (see Section 5, andRosner, 2000, Chapter 6) and characteristic time correlations is routinely used by engineers to design such complex physicochemical systems (see, e.g., Derr and Mellor, 1990 and Mattingly et al., 2002). In this complementary spirit we offer the present "zero-dimensional" model of a well-mixed spray combustor (i.e., "heterogeneous stirred reactor"). Even in its simplest (present) form, in many respects it transcends the capabilities of some earlier models (which, e.g., were "silent" about the effects of feed fuel spray polydispersity, neglected the effects of back-mixing even in the vapor phase, and/or assumed *quasi-steady*(QS) evaporation-rate behavior for each isolated droplet—see, e.g., Courtney, 1960—an assumption long known to be inaccurate at rocket motor pressures and not even valid for modern gas turbines). In Section 5we indicate how certain enhancements to this basic model can be systematically added, including, of course, consideration of downstream "modules". However, our main goal here is to illustrate what interesting information even this one-module, perfectly well-mixed non-QS vaporization rate-controlled model can already provide—including "details" of the droplet size distribution (DSD) emerging from the so-called "primary (two-phase) combustion" zone, and useful expressions for the: (a) vaporization-controlled rate of chemical energy release per unit chamber volume ("combustion *intensity*") and (b) corresponding combustion *efficiency* achievable in such "non-plug-flow" jet-stirred two-phase devices.

That these results and insights, directly linking spray combustor performance to injector characteristics, can be extracted in seconds using only repeated numerical quadratures (see 3.8 and 4) are both noteworthy and appealing, both for engineering and pedagogical purposes. Inevitably, our present conceptual and numerical illustrations (which emphasize only a kerosene-like fuel burning in an O_2-containing gas at pressures up to ca. 24 atm) are necessarily incomplete. However, these illustrations will hopefully suggest tractable applications or extensions of particular interest to the reader. Indeed, if the ratio of information gained to computational effort expended is any indication of the merit of a complementary approach, the present one appears to have much to recommend it.

The following short account of our theoretical approach and illustrative results is structured as follows. Our underlying idealizations and basic assumptions are first set forth in Section 2. This provides the basis for the theoretical development outlined in Section 3— which includes a summary of our non-QS treatment of individual droplet vaporization (Rosner and Chang, 1973 and Crespo and Liñán, 1975), our formally exact Eulerian-based population balance method for dealing with the evolution of a prescribed injected spray, and our simple method for coupling the vaporizing fuel spray with the "surrounding", combusting gas phase. Here we identify and compute two dimensionless performance indices of special importance, viz. the loss in efficiency associated with the fraction, f_{UNevap}, of the original liquid spray which remains UNevaporated, and the volumetric energy release rate associated with steady-, adiabatic- operation. All of our parametric studies and numerical illustrations are collected in Section 4, including typical fuel property parameters used (Section 4.1), and how the abovementioned combustor performance indices respond to changes in reference values of a Sauter-mean-diameter (SMD)-vaporization-based Damköhler number, our dimensionless measure of non-QS effects, and the injected fuel spray polydispersity parameter at each of several pressures. We return to our principal underlying assumptions in Section 5, defending some and relaxing or delimiting the validity of others. Some immediate and some less-obvious implications and extensions are also indicated there. Section 6, which completes this paper, recapitulates the essential features of this idealized spray combustor approach in the light of our admittedly ambitious objectives, and summarizes our principal conclusions and recommendations.

MATHEMATICAL MODEL; IDEALIZED SPRAY "PRIMARY" COMBUSTOR

Basic Assumptions Concerning Contacting Pattern, Individual Droplet Behavior, Rate-Controlling Steps, and Combustor Heat Loss

Considering the need for economy and tractability, we explicitly treat below a single-module steady-flow spray combustor, making the following principal basic assumptions/idealizations. (A brief discussion of their validity-limits and implications is postponed to 4 and 5. Clearly, we are deliberately sweeping aside many equipment-specific "details", and some of our assumptions will be more easily relaxed than others.) Inlet stream conditions and combustor pressure level are imagined to be fully specified—including the fuel injector-produced DSD (see Section 3.2).

A1 The vessel, of effective volume V_{ch}, is "well-(macro-) mixed", both with respect to the continuous (oxidizer- and combustion product containing gas-) phase and the fuel droplet (spray-) phase.

A2 The droplet-phase volume fraction, φ, is small enough to neglect droplet–droplet interactions in the prevailing mean gas mixture field (see, e.g., Labowsky, 1976, Labowsky, 1980 and Sangiovanni and Labowsky, 1982).

A3 Each droplet evaporates at the physical diffusion-controlled rate appropriate to the prevailing mean environment, its residence time in the vessel, and its initial diameter (see, e.g., Rosner and Chang, 1973 and Crespo and Liñán, 1975).

A4 The vapor-phase chemical heat release rate is controlled exclusively by the availability of fuel vapor and is spatially uniform (not localized near individual droplets).

A5 Time-lags associated with droplet heat-up (HU) and/or vapor-phase "ignition delay" are negligible compared to the total prevailing droplet vaporization lifetime, t_{vap}.

A6Neither droplets nor energy are appreciably lost to the confining vessel walls.

Regarding our first assumption, while the "well-stirred" reactors pictured in most ChE textbooks contain motor-driven physical impellers, as well as baffles fixed to the vessel inner wall, the "well-mixed" spray combustors we are considering here should be thought of as turbulent "jet-stirred" vessels. Thus, in high-intensity AGT and rocket combustors the "stirring" which leads to rapid molecular mixing is the result of the steady input of mechanical kinetic energy via the injected (often swirling) fluids. A significant portion of this kinetic energy is continuously converted to *turbulence* kinetic energy via jet entrainment, intense recirculation, fluid parcel distortion, and shear layer instabilities. Early atmospheric pressure radio-active tracer experiments of Bartok et al. (1960) suggested that, using "jet-stirring" (rather than physical impellers), some 85–95% of the total physical volume can often be considered "perfectly well-mixed". This is the effective volume, V_{ch}, which appears in the equations/parameters below. In this volume, all time-averaged quantities (associated with either the gas phase or the liquid phase) are spatially and temporally uniform, however, the local mixture fraction ^1Z will exhibit spatial and temporal "fluctuations" (often quantified by the rms variance $\langle Z'Z' \rangle^{1/2}$ compared to $\langle Z \rangle$) reflecting the incompleteness of molecular level ("micro-") mixing. Moreover, in this well ("macro-") mixed limit, the exit stream exhibits all of these attributes, without further change.

By exploiting these particular assumptions, together with well-established conservation (mass- and energy-balance-) conditions, our goal is to predict the self-consistent conditions existing in/exiting from such a well-mixed chamber. This includes the emerging fuel DSD, and the associated gas-phase temperature. These are used to define the "efficiency" of the spray combustor,[2] whose presumed purpose is to convert the chemical energy content of the initially liquid fuel into thermal energy of the predominantly gaseous combustion products.

Two dimensionless parameters will be seen to play an important role in the following idealized performance analysis. One is a characteristic time ratio (called here Dam_{QS}) defined as the ratio of a fluid parcel mean residence time $\langle t_{flow} \rangle$ in the combustion chamber, to the QS value of the vaporization lifetime of a droplet with the SMD of the feed spray. The second is a dimensionless parameter (called ε) governing systematic departures from the familiar QS-rate law for droplet vaporization. We remark that the QS presumption has been used in *all* previous simulations of spray combustor performance (see Section 3.1). In these terms it will also be possible to calculate the prevailing vaporization-controlled rate of chemical energy release per unit chamber volume—sometimes simply called the "combustion intensity" (or combustor "loading factor") and expressed in the engineering units: MW/m^3 or, for higher pressure chemical rockets, GW/m^3. An attractive feature of the present model will be seen to be the ease of studying the sensitivity of these measures of spray combustor performance to fuel injector ("atomizer"-) characteristics (e.g., SMD and DSD spread) and departures from QS vaporization rate conditions. While we focus below on the "primary" combustor module itself, we are also obtaining information essential to design any downstream modules—including, perhaps, a plug-flow "burn-out" module (see, e.g., Mattingly et al., 2002 and the brief summary of such two-module models in Rosner, 2000).

THEORETICAL DEVELOPMENT

Non-QS Area vs. Time Relation for an Individual fuel Droplet

Previous work on spray evolution in engines has been based on the so-called QS approximation for each individual droplet of radius R(t), according to which -dR/dt is simply proportional to 1/R, with a proportionality factor that is constant in a constant environment if the droplet is at the so-called "wet-bulb" temperature (often near—but systematically below $T_{bp}(p)$, This leads to a familiar result, referred to as the "d^2-law" that in such an environment the individual droplet *surface area will* decrease linearly with (residence-) time. As a corollary, the total QS diffusion-controlled droplet lifetime, quadratic in the initial diameter $_{d0}$, is found to be

$$t_{vap,QS} = \frac{\rho_L d_0^2}{8\rho_G D \ln(1 + B_m)}$$

(1)

(see, e.g., Law, 2006 and Rosner, 2000) where D is a representative mean fuel vapor Fick diffusivity in the hot gas mixture, and the dimensionless mass transfer driving force parameter B_m is given in terms of local fuel vapor mass fractions ω_v by

$$B_m = \frac{\omega_{v,w} - \omega_{v,ch}}{1 - \omega_{v,w}}$$

(2)

However, especially since the studies of Rosner and Epstein (1970), Duda and Vrentas (1971), Rosner and Chang (1973) and Crespo and Liñán (1975), it is known that this QS approximation fails whenever the following dimensionless parameter is not small:

$$\varepsilon \equiv \left(\frac{2}{\pi} \cdot \frac{\rho_G}{\rho_L} \cdot \ln(1 + B)\right)^{1/2}$$

(3)

as is the case for low boiling point fuels vaporizing in high-pressure environments. Indeed, in what follows we will make extensive use of the following two dimensionless functions, convenient approximations to which follow from the analytical asymptotic results of Crespo and Liñán (1975):

$$\frac{a}{a_0} = \left(\frac{v}{v_0}\right)^{2/3} = \text{Fct}_1\left(\frac{t}{t_{vap}(\varepsilon)}\right) \equiv F_1(\theta, \varepsilon)$$

(4)

and

$$\frac{t_{vap}(\varepsilon)}{t_{vap,QS}} \equiv F_2(\varepsilon)$$

(5)

A closed-form analytic approximation to $_{F1}(\theta,\varepsilon)$, probably acceptable for ε-values not much larger than 0.6, is (see Crespo and Liñán, 1975, Eq. 33)

$$F_1(\theta, \varepsilon) = 1 - \theta' - \varepsilon \left(\sqrt{\theta'} - \frac{1-\theta'}{2} \ln \frac{1-\sqrt{\theta'}}{1+\sqrt{\theta'}} \right)$$

(6)

where

$$\theta' \equiv F_2(\varepsilon)\theta \tag{7}$$

As observed in Fig. 1(a), $a/_{a0}$ vs. $t/t_{vap'QS}$ is non-linear (concave upward) whenever ε is non-zero. Fig. 1(b) reveals that $_{F2}$ is only initially near $(1+\varepsilon)^{-1}$, when ε is sufficiently small. Thus, above ca. 0.1 the magnitude of ε itself underestimates the relative error of applying the QS approximation to the vaporization lifetime of individual droplets in a population. It should be kept in mind that droplet vaporization in a finite length of time requires that ε be non-zero, since, using our definition of ε, Eq. (3) can be rewritten in the instructive form

$$t_{vap,QS} = \frac{1}{4\pi\varepsilon^2} \cdot \frac{d_0^2}{D}$$

(8)

It will be seen (see Section 4, and also in Rosner, 2006) that noticeable distortions in the steady-state DSDs emerging from a well-stirred vessel are associated with non-negligible ε-values—especially when the feed DSD has a narrow spread (see Section 3.2). The evaluation of self-consistent steady-state values of ε will be postponed to Section 3.5. We remark that in our present simulations (kerosene-like fuel up to p=24 atm) we encountered values of ε up to ca. 0.3. Interestingly, the constant property numerical calculations of Duda and Vrentas (1971) provide non-QS lifetime information up to ε-values of about 2.1.

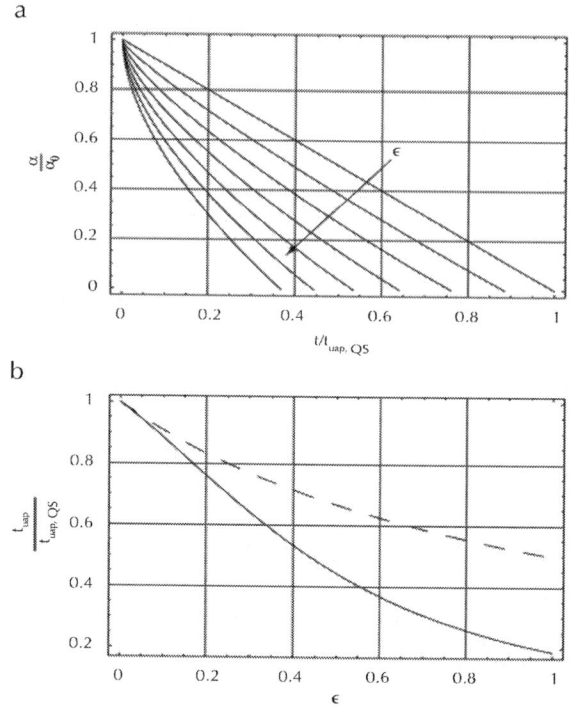

Figure 1: Normalized surface area vs. time "history" for the non-quasi-steady evaporation of an isothermal, isolated, spherical droplet in a constant environment. Based on (4) and (6) for ε=0, 0.1, 0.2, 0.3, 0.4, 0.5 and 0.6 (a). Corresponding vaporization time over the QS result as a function of ε (b). The dashed line shows the approximate result t_{vap}/t_{vap},QS~1/(1+ε).

As mentioned above, it appears that even the most elaborate spray combustor models investigated to date have relied on the QS "d²-law" to estimate the rate of fuel vapor release in the chamber. Accordingly, one valuable result of the present study is a direct comparison of predicted spray combustor performance using the more realistic (especially at high pressures) non-QS-rate law instead of the familiar QS droplet evaporation-rate law.

"Atomizer" DSD

Many continuous distribution functions have been used to describe the performance of liquid fuel injectors (see, e.g., Bayvel and Orzechowski,

1993 and Lefebvre, 1999). In any case, two essential parameters are a reference droplet diameter (which we will ultimately choose to be d_{32}=SMD) and a single dimensionless parameter which defines the DSD-breadth, "dispersity", or "spread". In what follows we will explicitly work in terms of number density distribution functions, $n(v)$, with respect to the size variable: $v=(\varpi/6)d^3$ defined such that

$$N_p = \int_0^\infty n(v)\,dv \quad \text{(0th moment of } n\text{)}$$

(9)

$$\varphi_p = \int_0^\infty v\,n(v)\,dv \quad \text{(1st moment of } n\text{)}$$

(10)

where N_p and φ_p are, respectively, the total droplet number density, and droplet volume fraction (assumed small compared to unity, see A2). The ratio of these two volume moments of $n(v)$ defines another convenient (number-average) reference size: $\bar{v} \equiv \psi_p/N_p$—indeed, our computed exit DSDs will be shown in the dimensionless/normalized form $\Psi(\eta)$, where $\psi \equiv \bar{v}\,n(v)/N_p$ and $\eta \equiv v/\bar{v}$.

Our spray combustor numerical illustrations (Section 4) will pertain to fuel injector DSDs well-described by a so-called "log-normal" (LN-) distribution function—for which $vn(v)$ is Gaussian in terms of $\ln v$ (rather than v itself). With the usual choice of parameters $(v_g,\ \sigma_g)$ this may be written in the normalized[3] form (see, e.g., Rosner and Tassopoulos, 1989, or Rosner and Khalil, 2000):

$$\frac{vn(v)}{N_p} = \frac{1}{\sqrt{2\pi}\ln\sigma_g} \cdot \exp\left(-\frac{(\ln v - \ln v_g)^2}{2\ln^2\sigma_g}\right)$$

(11)

From their definitions it is then easily shown that

$$\bar{v} = v_g \cdot \exp(\tfrac{1}{2}\ln^2\sigma_g)$$

(12)

and[4]

$$\frac{\pi}{6}\mathrm{SMD}^3 = v_g \cdot \exp\left(\frac{5}{6}\ln^2 \sigma_g\right)$$

(13)

Note that in this LN formulation a σ_g of unity corresponds to a Dirac-like DSD (mono-sized) spray. Actual injectors might produce sprays with σ_g-values somewhere in the range: $1.7 < \sigma_g < 2.3$. The sensitivity of idealized spray combustor performance to the values of both injector-produced SMD and the spread σ_g will be results of considerable theoretical and practical interest (Section 4). Our "baseline" calculations will be carried out for the intermediate value $\sigma_g = 2$ (see Fig. 2, Fig. 3, Fig. 4, Fig. 7 and Fig. 8 in Section 4).

Predicted Exit DSD and Its Moments for A Dirac-Fed Combustor (Role as Green's Fct.)

Our strategy for finding the steady-state DSD existing everywhere within (and emerging from) a well-stirred spray-fed combustor in the presence of non-QS, diffusion-controlled individual droplet evaporation is to first find an expression for the dimensionless DSD when the liquid feed is mono-sized, with droplet volume v_0 (this section). Then, in the absence of an appreciable frequency of binary droplet encounters leading to coalescence (or break-up (BU)[5]) within the jet-stirred vessel (A2), this function will serve as a *Green's function* to solve the more general problem of a combustor injected with, say, a LN DSD (Section 3.4)—i.e., a prescribed "spectrum" of v_0 values.

Setting aside, for the moment, the appropriate value of the non-QS parameter , we first solve the PBE for this simple special case, imagining that all droplets enter at the single size $_{v0}$. The steady-state balance between evaporation and the net inflow rate of droplets by convection then provides the following linear first-order ODE for n(v):

$$\frac{n - n_{\mathrm{Feed}}}{t_{\mathrm{flow}}} + \frac{\mathrm{d}}{\mathrm{d}v}[\dot{v}n(v)] = 0$$

(14)

where we have already introduced the mean residence time, t_{flow}, in the chamber of volume V_{ch}; i.e., $V_{ch}/[(\dot{m}_O + f_{vap} \dot{m}_F)/\rho_{ch}]$. If each spherical droplet loses surface area in accord with the diffusion-controlled rate law defined by (4) and (6) then, we can write

$$-\dot{v} = \frac{v_0}{t_{p,life}} \cdot \frac{3}{2} \cdot \left(\frac{v}{v_0}\right)^{1/3} \cdot \left(-\frac{dF_1}{d\theta}\right)$$

(15)

and

$$n_{Feed} = N_{p'Feed} \delta(v - v_0)$$

(16)

where $\delta(v-v_0)$ is the Dirac function of argument $(v-v_0)$, and $\theta \equiv t/t_{vap}(\varepsilon)$ is uniquely related to v/v_0 by inverting $(v/v_0)^{2/3} = F_1(\theta, \varepsilon)$ (Eq. (4)).

By integrating Eq. (14) we can then derive the following formally exact result for the dimensionless DSD: $G(v,v_0) \equiv v_0 n(v)/N_p$ emerging from (and existing within) the well-stirred vessel—i.e.:

$$G(v, v_0) = \frac{\frac{2}{3} e^{-\theta/Dam_0}}{Dam_0(1 - e^{-1/Dam_0})F_1^{1/2}(-dF_1/d\theta)}$$

(17)

where

$$Dam_0 = \frac{Dam_{QS,0}}{F_2(\varepsilon)}$$

(18)

and the dimensionless parameter $Dam_{QS,0}$, defined by

$$Dam_{QS,0} = \frac{t_{flow}}{t_{vap,QS,0}}$$

(19)

is the corresponding vaporization-based Damköhler number for a droplet of initial size v_0 vaporizing following the QS vaporization rate law.

Several properties of this DSD were studied and discussed in Rosner (2006), where significant effects of non-zero ε were found, especially for v/v0 near unity and ε-values above 0.1. It is also possible to track the dimensionless volume moments of G—for example, we find

$$\mu_k = \frac{\int_0^1 F_1(\theta, \varepsilon)^{3k/2} e^{-\theta/\text{Dam}} \, d\theta}{\text{Dam}(1 - e^{-1/\text{Dam}})} \tag{20}$$

But, in the present context, exploiting linearity (a consequence of our Assumption 2), $N_{p,1}G(v,v_0)/v_0$ will serve as a Green's function to provide (Section 3.4, Eq. (21)) the DSD for the more practical case of a chamber fed with an *arbitrary* DSD—in our present examples: LN.

Exit DSD for a "Polydispersed" Spray-Fed Combustor

If the fuel injectors provide a spray with a DSD given by nFeed(v_0), then the outlet DSD can be formally written in the *convolution* form

$$n(v; \varepsilon, \text{Dam}_{QS}) = \int_0^\infty \frac{n_{\text{Feed}}(v_0)}{N_0} \cdot \frac{N_1 G(v, v_0; \text{Dam}_0)}{v_0} \, dv_0 \tag{21}$$

Where

$$\frac{N_1}{N_0} = 1 - \exp\left(\frac{-1}{\text{Dam}_0}\right) \tag{22}$$

In this process we must also keep in mind that the previously defined "parameter" Dam0 also contains the size "parameter" v_0. For our present (polydispersed spray-) case we will define a *system* Dam based on a QS-evaporating droplet with the particular (Sauter-) mean diameter: SMD=d_{32}, where the corresponding volume is related to vg by Eq. (13). On this basis, if we specify the values of the parameters Dam_0, based on SMD, ε (see, also, Section 3.5), and the remaining injector spray parameter (say, σ_g), it is possible to calculate by quadratures alone, the exit DSD, its dimensionless equivalent: $\psi(v / \bar{v})$, and its dimensionless moments, μ_k. (Particular values of k have special physical significance, as briefly discussed below, and more extensively in Rosner, 1989, Rosner and Tassopoulos, 1989 and Rosner et al., 2003.) Of particular interest here will be the volume fraction of the exit spray,—i.e., M1. This quantifies the fraction of the initial spray which "survives" the evaporation process—i.e., remains un-evaporated, and, hence, determines the efficiency of the one-module ("primary zone") combustor (see Section 3.6). As shown in Section 3.7, this is also a sufficient level of information to calculate the prevailing vaporization rate-controlled chemical energy release rate per unit chamber volume—often simply called the combustion "intensity". Indeed, it will be interesting to compare these calculated values not only to those computed using a QS-evaporation-rate law, but also those actually reported for high-performance gas turbine combustors, and higher pressure chemical rockets—as discussed further in Section 5.

While not treated here, we conclude this section by remarking that these DSD results would also provide the input data for the analysis of any downstream modules of design interest—e.g., a plug-flow "burn-out" module (within which the non-QS parameter ε would no longer be spatially uniform).

Balance Equations Coupling the Gas Phase with the Spray

For the well-stirred vessel (module) we return here to the important question of what determines the self-consistent, spatially uniform value of the non-QS parameter ε. This value is not known in advance since it must be consistent with the heat release associated with the actual amount of fuel vaporized. The same can be said of the prevailing value of Dam_{QS}, which will systematically differ from $Dam_{QS,ref}$ (see below). To illustrate this as simply as possible we introduce here two further assumptions concerning gas mixture properties, less essential than those enumerated in Section 2. One is that of near-equality of the relevant gas mixture molecular diffusivities, so that the abovementioned Fick diffusivity D is not very different from the thermal diffusivity, $k/(\rho_{cp})$

of the prevailing gas mixture (ie. $Le \approx 1$). The second assumption deals with the use of a suitable mean heat capacity, written $\langle c_p \rangle$ products, to simplify the enthalpy balance. If one accepts these simplifications, relating the actual rate of heat generation (via fuel vapor combustion) to the enthalpy rise of the combustion product mixture leads to a steady-state compatibility condition of the algebraic form

$$T_{ch} - T_{Feed} \simeq \frac{f_{vap} \Phi f_{stoich}}{1 + f_{vap} \Phi f_{stoich}} \cdot \frac{Q_{comb}}{\langle c_p \rangle_{products}} \tag{23}$$

when $\Phi < 1$, and, when $\Phi > 1$:

$$T_{ch} - T_{Feed} \simeq \frac{1}{\Phi} \cdot \frac{f_{vap} \Phi f_{stoich}}{1 + f_{vap} \Phi f_{stoich}} \cdot \frac{Q_{comb}}{\langle c_p \rangle_{products}} \tag{24}$$

where Qcomb is the "lower" heat of combustion of the liquid fuel (see Section 4.1). The indicated mixture temperature, T_{ch}, not only determines the gas density appearing in the non-QS parameter ε (via the mixture EOS, which need not be that of an ideal gas, see, e.g., Rosner and Arias-Zugasti, 2007), it also fixes the value of the vaporization "driving force" parameter Bh, which is approximately given by

$$B_h = \frac{\langle c_p \rangle (T_{ch} - T_{WB}(p))}{L_{vap}(T_{WB}(p))} \tag{25}$$

(see. e.g., Rosner, 1972 and Rosner, 2000). Note that since we will assume $T_{Feed} = T_{WB}(p)$, the numerator of B_h is calculated using (23) and (24).

Turning to the prevailing value of Dam_{QS} we find from their definitions that

$$Dam_{QS} = \frac{1 + \Phi f_{stoich}}{1 + f_{vap} \Phi \cdot f_{stoich}} \cdot \frac{(\rho D)_g}{(\rho D)_{g,ref}} \cdot \left(\frac{\varepsilon}{\varepsilon_{ref}} \right)^2 \cdot Dam_{QS.ref} \tag{26}$$

We see that the values of Dam_{QS}, f_{vap}, ηcomb, and ε must be self-consistent and will, in general, differ systematically from their respective (ref) values: $\text{Dam}_{QS,ref}$, $\exp(-1/\text{Dam}_{QS,ref})$, $\exp(-1/\text{Dam}_{QS},\text{ref})$, and ε_{ref}. In a procedure investigated below (Section 3.8), we specify "nominal" (reference) values of the parameter ε (using the adiabatic condition for complete fuel vaporization and combustion), and then iteratively converge on self-consistent DSD, ε-values, from which all other quantities of interest (including the corresponding steady-state combustion intensity and efficiency) are computed.

Combustion Efficiency and Fuel Fraction Evaporated

As mentioned in Section 2.1, to the present level of approximation, when $\Phi<1$ the efficiency of this idealized spray combustor module, η_{comb}, may be identified with the fraction of the fuel spray injected which is successfully vaporized—i.e., $1-f_{UNevap}$. Here f_{UNevap}, the fuel fraction escaping evaporation (or "unevaporated") can be calculated from both the feed and chamber exit spray volume fractions:

$$f_{UNevap} = 1 - f_{vap} = \frac{\varphi_{ch}}{\varphi_{injector}} = \frac{M_{1,ch}}{M_{1,F}} \tag{27}$$

Where M1 is the 1st volume moment of the appropriate DSD: n(v). When $\Phi>1$ our definition of η_{comb} leads to the corresponding result

$$\eta_{comb} = \frac{1 + \Phi f_{stoich}}{1 + f_{vap}\Phi f_{stoich}} \cdot f_{vap} \tag{28}$$

We note here that the effective feed spray volume fraction entering this equation must be compatible with the prescribed value of the *combustor equivalence ratio* , Φ, as well as being corrected to the gas density, ρ_{ch}, existing in the chamber during steady-state combustion. On this basis we find

$$\varphi_{\text{Feed,eff}} = \frac{\rho_{\text{ch}} f_{\text{stoich}} \Phi}{\rho_p - (\rho_{0,\text{Feed}} - \rho_{\text{ch}}) f_{\text{stoich}} \Phi}$$

(29)

where ρ_p is the liquid fuel mass density at the feed temperature, and $\rho_{0,\text{Feed}}$ is the vaporized oxidizer density at chamber pressure p and feed temperature, T_{Feed}. Note, also, that once $\varphi_{\text{Feed,eff}}$, the feed Sauter-mean droplet diameter (or volume), and DSD spread σ_g are specified, this defines $N_{p,\text{Feed,eff}} = M_{0,\text{Feed,eff}}$—completing the present description of the feed spray DSD. Calculation of the combustion efficiency via (27) and (28) then proceeds by evaluating the first moment of the steady state-chamber/exit DSD, $M_{1,\text{ch}}$, calculated via our abovementioned "convolution" procedure (Section 3.4). Of course, the fraction evaporated, fvap, explicitly enters the energy balance, (23) and (24), needed to calculate the steady-state value of our non-QS parameter ε, as well as the prevailing Dam, again revealing the level of coupling between the liquid fuel spray and the surrounding combustion product (plus excess oxidizer) gaseous mixture.

An interesting generalization that allows for the fact that not all of the fuel vaporized may fully react chemically (even in the presence of excess oxidizer), along with its implications for the prediction of "combustor flame-out", is being studied at present time, and it will be presented in a forthcoming publication.

Vaporization-Controlled Volumetric Heat Release Rate

Using iterative methods, we will calculate (Section 3.6) the idealized spray combustor efficiency associated with each specified pair of parameters, $\text{Dam}_{\text{QS,ref}}$ and ε_{ref}, where the reference condition is that specified earlier. Then, since the actual chemical energy release rate per unit chamber volume can be written (in the presence of excess oxidizer, i.e., mixture ratio $\Phi < 1$):

$$\langle \dot{q}'''_{\text{chem}} \rangle = \frac{\dot{m}_L Q_{\text{comb}}}{V_{\text{ch}}} \cdot f_{\text{vap}}$$

(30)

it follows from our definitions that the corresponding *combustion intensity* can be computed from

$$\langle \dot{q}_{chem}''' \rangle = \frac{4\pi(\rho D)_{g,ref} Q_{comb}}{SMD_{Feed}^2} \cdot \frac{\Phi f_{stoich}}{1 + \Phi f_{stoich}} \cdot \frac{\varepsilon_{ref}^2 f_{vap}}{Dam_{QS,ref}} \tag{31}$$

(When $\Phi > 1$ the RHS of Eq. (31) is merely modified by the presence of an additional factor: $1/\Phi$). Thus, a knowledge of the reference state and the associated values of f_{vap} and η_{comb} for the prescribed values of $Dam_{QS,ref}$ and propellant mixture ratio Φ is also sufficient to calculate the corresponding steady-state value of the vaporization-controlled *combustion intensity*. Indeed, in Fig. 2, Fig. 6, Fig. 7 and Fig. 8 we plot the product of the last two dimensionless factors appearing in Eq. (31). This is equivalent to a *dimensionless combustion intensity*, I, using the single droplet evaporation-rate term: $4\pi (pD)_{g,ref} Q_{comb}/SMD^2$ as an appropriate "reference" value. Indeed, to within a factor of $ln(1+B)$ (which does not change its order of magnitude) I is nearly equal to the ratio of $\langle q_{chem}''' \rangle$ in the chamber to $m_F Q_{comb}/V_d$ for a single evaporating droplet of volume $V_d = (\varpi/6)SMD^3$.

In this connection, it may be of interest to remark here that in the now-classic droplet vaporization-controlled study of Priem and Heidmann (1960) the corresponding combustion "intensity" is not actually explicitly discussed or reported. But, this notion is, of course, *implicit* in their calculations/correlations of required combustor lengths for acceptable efficiencies. Moreover, even the prominent role of our present vaporization-based Damköhler number parameter, DamQS, is implicit, as can be appreciated from their comment (Discussion and Conclusions, p. 49): "For high efficiency the combustor must therefore be designed with sufficient stay-time for the large drops to vaporize". However, the reader will appreciate that in our present well-mixed chamber there is inevitably a broad distribution of residence (stay-) times (see, e.g., Rosner, 2000) so that, in effect, all droplet sizes fed to the chamber will actually appear in the combustor effluent. Our population balance-based computational procedure (3.3 and 3.4) not only accounts for this distribution of residence times, but also the DSD-consequences of non-QS individual droplet evaporation (Section 3.1).

Computational Considerations

The value of this idealized spray combustor model rests, in part, upon our ability to rapidly make self-consistent spray combustor performance calculations including: f_{vap} and $_{\eta comb}$, I, output DSDs, and to associate them with convenient reference values of the parameters Dam_{QS}-and ε—all via a simple sequence of repeated numerical quadratures. This is demonstrated in Section 4.2, which includes further comments about our procedures, along with graphs (Fig. 2, Fig. 3, Fig. 4, Fig. 5, Fig. 6, Fig. 7 and Fig. 8) of our representative numerical results for f_{vap}, η_{comb}, I, Ψ, μ_k presented in reasonably "universal" form and covering a wide range of parameter-values: p, Φ, Dam, σ_g.

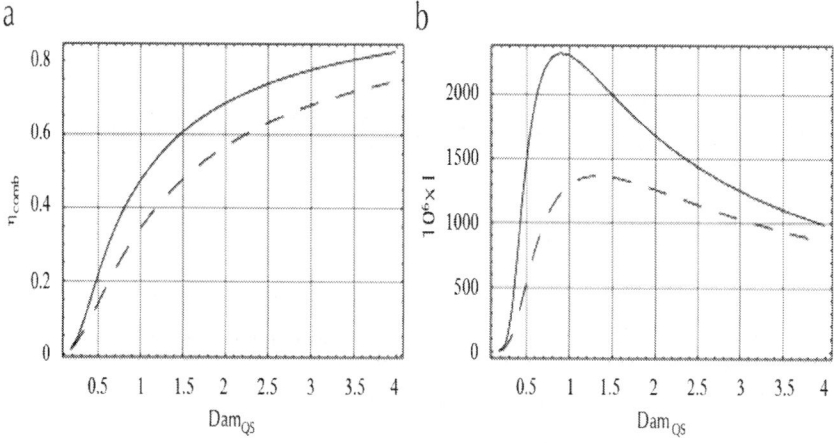

Figure 2: Predicted Dam_{QS} dependence of the combustion "efficiency" η_{comb} (a) and dimensionless combustion intensity I (b) for the ("base"-) case: P=20atm, σ_g=2, Φ=0.8. The results shown in solid lines have been computed using the non-quasi-steady vaporization rate law Eq. (6), the dashed lines have been computed using the corresponding quasi-steady vaporization rate law found by assuming ε=0 in Eq. (6).

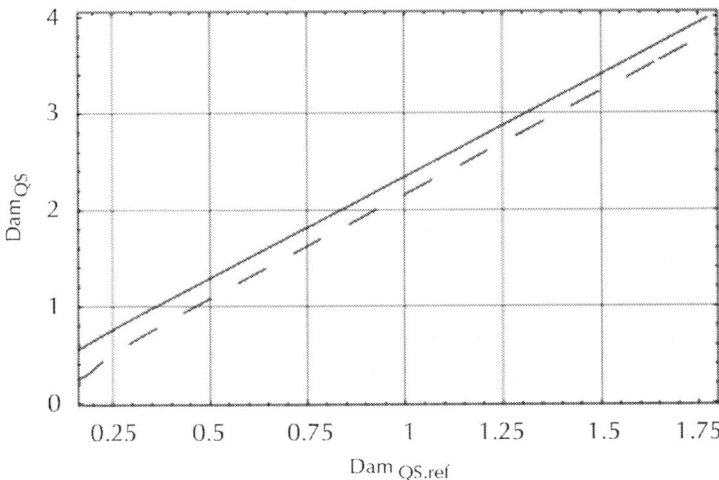

Figure 3: Relation between Dam_{QS} and $Dam_{QS,ref}$ for the ("base"-) case: P=20atm, $\sigma_g=2$, $\Phi=0.8$. The solid line corresponds to the non-QS-rate law Eq. (6), the dashed line to the QS-rate law ($\varepsilon=0$).

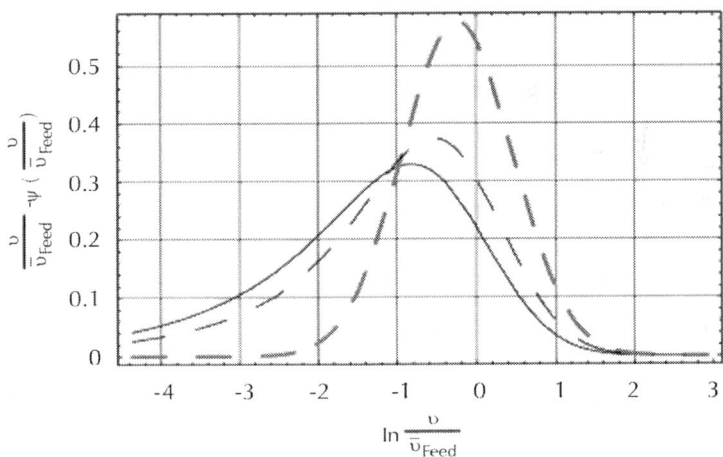

Figure 4: Comparison of predicted output and input (thick, dashed, gray line) droplet size distributions showing the non-quasi-steady evaporation-rate effect for the ("base"-) case: P=20atm, $\sigma_g=2$, $\Phi=0.8$, $Dam_{QS}=1$. The solid line corresponds to the non-QS-rate law Eq. (6), the dashed line to the QS-rate law ($\varepsilon=0$).

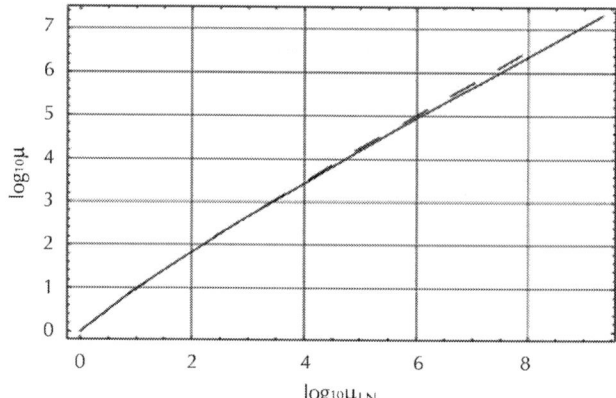

Figure 5: Dimensionless moments (μ_k with k between 1 and 5) of the output DSD (with respect to $v/\bar{v}$, where $\bar{v}$ is the NMV of the output DSD), plotted vs. the corresponding moments of a log-normal DSD (departures from linearity betray non log-normal behavior). The results shown correspond to the ("base"-) case: P=20atm, σ_g=2, Φ=0.8 and Dam_{QS}=1. The solid line corresponds to the non-QS-rate law Eq. (6), the dashed line to the QS-rate law (ε=0).

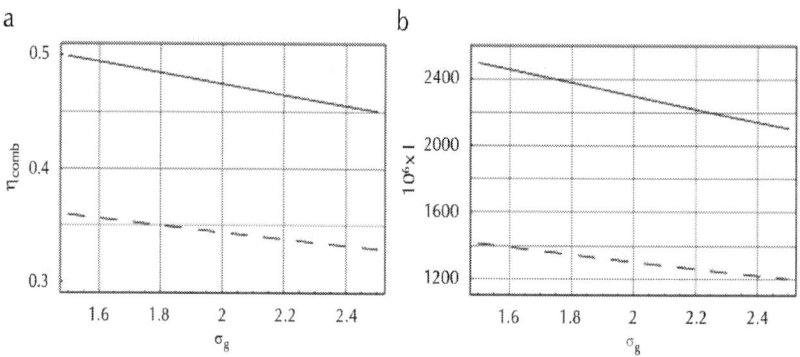

Figure 6: Predicted dependence of spray combustion "efficiency" η_{comb} (a) and dimensionless combustion intensity I (b) on the feed DSD spread σ_g for the case:P=20atm, Φ=0.8 and Dam_{QS}=1. The solid lines have been computed using the non-quasi-steady vaporization rate law Eq. (6), the dashed lines using the corresponding quasi-steady vaporization rate law (ε=0).

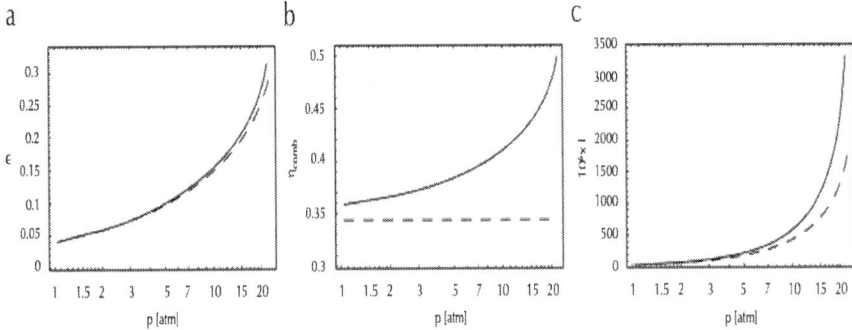

Figure 7: Predicted pressure dependence of the non-QS parameter (a), combustor "efficiency" η_{comb} (b) and dimensionless combustion intensity I (c) for the case: $\sigma_g=2$, $=0.8$ and $Dam_{QS}=1$. The solid lines have been computed using the non-quasi-steady vaporization rate law Eq. (6), the dashed lines using the corresponding quasi-steady vaporization rate law ($=0$).

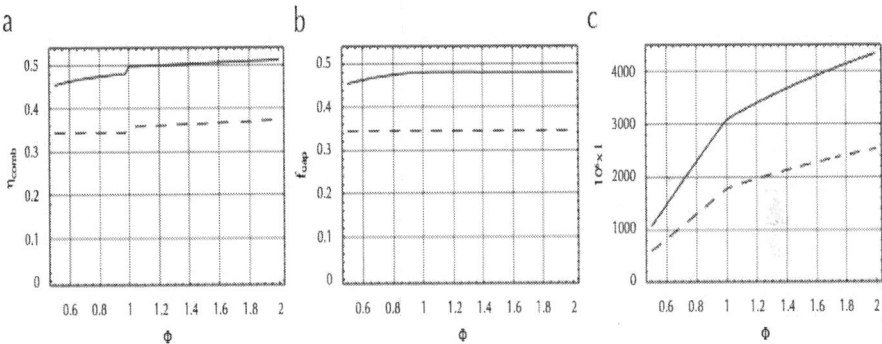

Figure 8: Predicted dependence of combustor "efficiency" η_{comb} (a), fuel fraction evaporated fvap (b) and dimensionless combustion intensity I (c) on the overall equivalence ratio Φ for the case: P=20atm, $\sigma_g=2$ and $Dam_{QS}=1$. The solid lines have been computed using the non-quasi-steady vaporization rate law Eq. (6), the dashed lines using the corresponding quasi-steady vaporization rate law ($\varepsilon=0$).

One of our goals is to compute the actual shape of the exit DSD function, and compare it to the input LN distribution—i.e., to show that, in general, the presumption of an invariant DSD-*shape*   is too crude. To that end the exit DSD has been approximated by means of a spectral expansion. In order to achieve high accuracy, even at very low values of the variable v, the DSD function has been computed in terms of the variable lnv, accordingly the spectral expansion used was based on the Hermite basis (lnv∈[-∞,∞]), the method used to calculate the coefficients of the spectral expansion has been Orthogonal Collocation—for a review on spectral methods in general, and on the Orthogonal Collocation method in particular, see e.g., Boyd (2000). In all the cases considered, the corresponding exit DSD has been computed using Eq. (21) to calculate the values of the DSD at the corresponding abscissae, which provide the corresponding coefficients for the spectral expansion. An analysis of the convergence rate of such expansion was performed to determine the number of spectral components needed. According to that analysis the convergence rate of the spectral expansion depends mainly on the Damköhler number, to a lesser extent on the non-QS parameter ε, and it was almost independent of the feed DSD spread σ_g (for σ_g-values in the range of physical interest $1.7 < \sigma_g < 2.3$). The lowest convergence rates found correspond to large values of Dam. For instance, about 40 spectral components are needed for an accurate spectral expansion forDam=3, while, for values of Dam<1, 20 spectral components were enough to achieve convergence. The reason for this slow convergence rate at high Dam is because when Dam>1 the exit DSD becomes very broad (in terms of lnv), forcing the inclusion of a relatively large number of abscissae to cover the whole region where the DSD is appreciably different from zero.

Based on this analysis, and also on the fact that the needed calculations could be performed very fast,[6] 48 spectral components have been used in all the cases shown, irrespective of the value of the vaporization Damköler number. Further implications of these methods/results will be discussed in Section 5.

ILLUSTRATIVE QUANTITATIVE RESULTS: PREDICTED PERFORMANCE OF IDEALIZED SPRAY COMBUSTORS

Fuel/Oxidizer System of Interest—typical Parameters

A fuel/oxidizer system of particular interest, historically owing to aircraft gas turbine and ramjet applications, is that of a kerosene-like liquid sprayed into the combustor of an "air-breathing" engine. Thus, the illustrative calculations below will be carried out using the liquid fuel properties assembled in Table 1, with the oxidizer being $O_2(g)$ contained in non-vitiated air. The numerical values listed will be seen to be rather typical of petroleum-based liquid fuels that go under such common names as: "kerosene", "JP-1","RP-1", "gas-oil" or even "light (or high-speed-)-diesel"—with a nominal stoichiometry not far from $(CH_{1.9})_n$.

Table 1: Nominal properties of typical jet propulsion-type liquid fuel

Thermo-physical/chemical property	Value	Units
Molecular weight (MF)	154	kg/kmol
Liquid density (ρL)	0.82	t/m3
Normal boiling point (Tnbp,eff)	520	K
Latent heat of vaporization (Lvap at Tnbp)	0.289	MJ/kg
Trouton ratio (ΔHvap/(RTnbp,eff))	10.3	dimensionless
Heat of combustion (Qcomb)	43.1	MJ/kg (liquid)
Stoichiometric F/air mass ratio (fstoich)	0.0681	dimensionless
Effective critical parameters and vapor pressure constantsa (when needed)		

Critical pressure (pc)	20.72	atm
Critical temperature (Tc)	660	K
Critical volume (Vc)	0.627	m3/kmol
Critical value of pV/(R T) (=Zc)	0.24	dimensionless
Acentric factor (ω)	0.5	dimensionless
Effective Lennard-Jones 12:6 intermolecular potential parameters (when needed)		
Molecular diameter (σ11)	0.837	nm
Energy well parameter (11/kB)	368	K
Antoine vapor pressure constantsa		
A	7.96	dimensionless
B	2031	K
C′	120	K

[a]Parameters in the vapor pressure curve-fit: log10psat(Torr)=A-B/(T-C').

Thus, in the pressure range between ca. 1 atm, and, say, 20.72 atm, it is possible to estimate the prevailing droplet boiling point temperature from an inverted Antoine-type equation of the form

$$T_{bp}(p) = 120 + \frac{T_{nbp} - 120}{1 - \ln(p(\text{atm})/11.695)}$$

$$(32)$$

However, due to evaporative cooling, the actual droplet temperature, T_{WB}, will be somewhat smaller than the prevailing fuel boiling point temperature. Indeed, a simple energy balance reveals that when $Le=1$ then T_{WB} will be the solution to the transcendental equation:

$$B_h(T_{WB})=B_m(T_{WB}) \tag{33}$$

where B_m is defined by Eq. (2). If we use a pseudo-binary mixture assumption and invoke the Antoine vapor pressure equation (Table 1) then an efficient iterative method to obtain T_{WB} can be used[7] Appearing on the RHS of this equation is the temperature-dependent latent heat, which will be calculated from the correlation

$$L_{vap}(T_{WB}(p)) = \left(\frac{T_c - T_{WB}}{T_c - T_{nbp}}\right)^{0.4} \cdot L_{vap}(T_{nbp})$$

(34)

The liquid fuel mass density will be required in order to estimate the non-QS parameter ε. For this purpose use can be made of the correlation

$$\rho_L = \rho_c \cdot Z_c^{-(1-(T_{WB}/T_c))^{0.2857}}$$

(35)

where, of course, $\rho_c = M/V_c$.

An interesting consequence of evaporative cooling is the fact that subcritical droplet temperatures are associated with supercritical chamber pressures (see, e.g., Rosner and Chang, 1973). But in the present case we obtain subcritical droplet temperatures only at pressures up to ca. 24 atm—not much above the nominal critical pressure (20.72 atm) of the fuel. Non-QS evaporation-rate effects (non-negligible ε-values) clearly become more prominent under such conditions, but more complex property evaluation algorithms (e.g., for $\langle c_p \rangle$; see brief discussion below) are also required for self-consistency.

Since our pressures will not exceed ca. 24 atm and chamber temperatures will usually exceed 1500 K, for present purposes we simply calculate gas densities in the chamber from the perfect gas law:

$$\rho_g = \frac{pM_{ch}}{\mathscr{R}T_{ch}}$$

(36)

where, in the illustrative cases under consideration, the mean molecular weight, M_{ch}, will not depart much from 29 kg/kmol when <1. For M_{ch} when >1

When needed, gaseous combustion product heat capacities will simply be calculated from

$$\langle c_p \rangle_{ch} \simeq 1.14 \cdot \left(\frac{T}{1000}\right)^{0.20} kJ/(kg\,K)$$

(37)

when the oxidizer is "air". While not included in Table 1, the unburned fuel vapor heat capacity will be represented by

$$\frac{k}{\rho c_p}(T = 1500\,\mathrm{K}, p = 1\,\mathrm{atm}) \simeq 3.28 \times 10^{-4}\,\mathrm{m^2/s}$$

$$(38)$$

Finally, molecular diffusivities, when needed, will be calculated from the near-stoichiometric estimate

$$\frac{k}{\rho c_p}(T = 1500\,\mathrm{K}, p = 1\,\mathrm{atm}) \simeq 3.28 \times 10^{-4}\,\mathrm{m^2/s}$$

$$(39)$$

scaled in accord with a $T^{1.75}/p$ (atm) ideal-gas kinetic theory dependence. While this procedure will overestimate D for dilute (very fuel-lean) mixtures, it will underestimate$D$ near the vaporizing droplet surfaces. Accordingly, this approximation will probably be sufficient for our present illustrative purposes. Clearly all of the above estimates, including approximate constancy of the effective heat of combustion (Table 1) on the "fuel-rich" side, could be readily improved with the use of more elaborate gas mixture property subroutines—incorporating even systematic departures from the ideal gas law.

Representative Performance Predictions: Well-mixed Adiabatic Spray Combustor

Using the property estimates of Section 4.1 (for the typical kerosene/air system) we have generated Fig. 2,Fig. 3, Fig. 4, Fig. 5, Fig. 6, Fig. 7 and Fig. 8 to illustrate what this predictive mathematical model can reveal about the dependence of spray combustor performance on the following basic parameters: SMD-based vaporization Damköhler number, pressure level, spread of the feed DSD, and overall equivalence ratio Φ(with $\Phi<1$ corresponding to the fuel-lean, or "excess air", regime). In each case we include corresponding results if one formally makes the ubiquitous QS approximation ($\varepsilon=0$)—which becomes increasingly unrealistic as one considers high pressure operation (see Figs. 7(a) and (b)).

Figs. 2(a) and (b) show the combustion efficiency and intensity, respectively, for our "base-case" calculation conditions of p=20 atm, σ_g=2, Φ=0.8. In these figures the solid lines represent the results predicted using non-QS droplet evaporation rate, while the dashed lines represent those predicted using the commonly used QS-rate law. Quite remarkable is the fact that while the combustion efficiency (equivalent to f_{vap} when <1) continues to rise with increasing Dam_{QS}, the combustion intensity passes through a maximum for Dam_{QS} values of order unity. Put another way: the volume added in order to vaporize an increasing fraction of the input fuel does not provide a proportionate increase in the rate of chemical energy release. Moreover, the predicted non-QS combustion intensity is about twice that expected using the QS-vaporization rate law. Also noteworthy is the dramatic increase in combustion intensity associated with higher pressure operation (Fig. 7(c)), and the significant increase associated with fuel-rich operation (Fig. 8(c)). The performance penalties associated with increased feed DSD spread are clearly shown in Figs. 6(a) and (b).

Representative information on how the DSD emerging from the chamber compares to the LN-feed DSD is contained in Fig. 4 and Fig. 5, which clearly show that the emerging DSD is NOT LN and is shifted to smaller diameters relative the QS prediction.

Of course, results for only a subset of the extensive collection of possible combinations have been collected here. As has been noted, for the "baseline" case we selected the particular combination: $Dam_{QS}=1$, p=20 atm, $\sigma_g=2$ and $\Phi=0.8$. Variations about this "baseline" case were then shown for each of the performance indices of interest: i.e., efficiency, fraction vaporized, combustion intensity, and the computed non-QS parameter ε defined earlier (Fig. 7(a)). The close relation between the reference- and prevailing- vaporization Damköhler number is displayed in Fig. 3.

Even without enhancements to further increase the predictive power of this relatively simple mathematical model (beyond the scope of this account, but see Section 5.2) our conclusion is that much can be learned from a study of these parametric dependencies. Further implications and extensions are considered in the light of our present underlying approximations in Section 5, which follows.

DISCUSSION: IMPLICATIONS, APPLICATIONS, EXTENSIONS

Historical Perspective

As discussed in greater detail in Sirignano (1999) and in a forthcoming review by one of us (DER), virtually all early theoretical models of spray combustors, including those which are rather computationally demanding, were based on a quasi-one-dimensional model which completely neglects back-mixing (or streamwise "dispersion")—even in the vapor phase (see, e.g., Probert, 1946, Shapiro and Erickson, 1956, Spalding, 1959, Priem and Heidmann, 1960, Williams, 1985, Nuruzzaman et al., 1971 and Sutton et al., 1972). Such "plug-flow" models are also often used to develop/test the performance of moment methods as applied to the population balance equation (see, e.g., Archambault et al., 2003 and Rosner et al., 2003). Yet, it was well-known that, at least locally, intense recirculation is needed to stabilize (anchor or achieve "self-piloted"-) combustion in such devices—which undoubtedly motivated Swithenbank etal. (1973) to explicitly introduce "modular" models of such combustors featuring perfectly well-mixed "primary" components. While the emphasis in that work, and its extensions, was on the role of complex vapor-phase chemical kinetics in establishing AGT-combustor "blow-out" limits and pollutant levels, we are here more concerned with elucidating, as simply as possible, the connection between liquid fuel injector ("atomizer") performance and spray combustor performance. Moreover, with regard to liquid propellant rocket motor *energy release rates per unit chamber volume*, it was already clear from the work of Bittker and Brokaw (1960) that vapor-phase (homogeneous) chemical kinetics was NOT the rate-limiting factor—i.e., actual values of $\langle q'''_{chem} \rangle$ were systematically lower than those expected from the intrinsic homogeneous chemical kinetics by well over three orders-of-magnitude. For spray-fed devices the plausibility and tractability of the alternative [8] of droplet vaporization rate control motivated the now-classical (plug-flow/QS vaporization) study of Priem and Heidmann (1960) and its many extensions, which have demonstrably proven useful to designers of liquid propellant rocket motors (see, e.g., Huzel and Huang, 1992). Indeed, these precedents encouraged us to revisit such a vaporization rate-controlled idealized spray combustor model, but this time adopting the much simpler "zero-dimensional" picture of the classical "well-mixed" vessel—as perhaps anticipated, but evidently not carried to fruition, by Courtney (1960) and, more recently, Lefebvre (1999). Moreover, from our recent "population-balance" research on the role of more accurate single-particle rate laws (Rosner, 2006), it also became clear that this program could be implemented even relaxing the ubiquitous QS individual droplet evaporation-rate assumption—as outlined in Section 3.1.

In displaying representative parametric results (Section 4), for convenience and generality we have here presumed an independence among parameters that may not exist in practice. Thus, for example, at any chamber pressure level actual injector performance may be such that the SMD and DSD spread of the fuel spray respond to the liquid feed flow rate and its viscosity—hence temperature. Conversely, we have here assumed $T_{Feed} \simeq T_{WB}(p)$. These kinds of "interdependencies" must be kept in mind in interpreting our present results/trends—especially for such applications as aircraft gas turbine combustors. However, they could, if desired, be readily "built-into" future more system-specific implementations/adaptations of such a model. That said, it is appropriate to investigate several of our underlying assumptions here—albeit in a preliminary fashion.

Examination/Relaxation of Underlying Assumptions

To achieve our stated objectives we have introduced many sweeping simplifications (A1–A6, Section 2)—especially the neglect of other relevant rate processes, including: droplet heat up, gas-phase ignition, turbulent mixing (for bi-propellant systems) and possibly even liquid-phase chemical heat release (e.g., for certain monopropellants or "hypergolic" bi-propellant combinations). We have also explicitly limited ourselves to the case of negligible loss of droplets and/or energy to the confinement walls of the combustion chamber. But even at this level, our present formulation, involving only iterative numerical quadratures, appears to economically provide useful information (Sections 4.2 and 4.3) about how spray combustor performance will depend on such factors as: injector ("atomizer") performance, available chamber volume, inlet (feed) conditions (liquid phase and gas phase), liquid fuel properties, and overall propellant mixture ratio.

Can The Gas-Phase Combustion Chemistry "Keep up"?

We have assumed that the gas-phase chemistry is capable of oxidizing the fuel vapor as fast as it can be supplied by the evaporating droplets (see A4 of Section 2.1). While a critical discussion of such complex homogeneous chemical kinetics is certainly beyond the scope of our present treatment, this presumed capability can be quantitatively assessed with the help of laminar premixed flame speed data/theory for the fuel vapor/oxidizer combination of interest. This generalization will also allow us to estimate both fuel-lean and fuel-rich spray combustor "flame-out". Interestingly enough, such considerations seem to indicate that the intrinsic homogeneous oxidation kinetics near the stoichiometric condition are capable of chemical heat release rates of ca. 3 TW/m^3 at 20 atm—some three orders of magnitude higher than those actually considered in Section 4.1!

Turbulent Enhancement of Fuel Droplet Evaporation Rate?

Another systematic effect which can be quantified within the framework of our present model is the role of combustion chamber *turbulence* in accelerating the time-averaged droplet vaporization rate. This generalization can be discussed using extensions of the single droplet work of Birouk and Gokalp (2006). Provided non-QS effects are themselves small, and one allows for the dynamical response of "untethered" droplets, such turbulence effects can be incorporated into the area vs. time relation for individual droplets (work in progress).

Further Generalizations

Another generalization of possible practical interest has to do with the consequences of injecting the liquid spray at a temperature appreciably[9] below the previously discussed adiabatic chamber "wet-bulb" temperature. In that case one must account for the *time* it takes each droplet to heat up to T_{WB}, as well as the *energy* required for this HU process. To bracket this effect one can imagine an instructive limiting case in which each droplet in the population heats up to T_{WB} with little loss of vapor (and perhaps even some thermal expansion!), and only then vaporizes as predicted by the present non-QS approach. From an energy point of view the sensible energy required to heat these droplets would therefore decrease the combustion-induced temperature rise, thereby lowering T_{ch} (Eq. (23)). We mention this as but one example of possible generalizations of some practical interest which would essentially preserve the simplicity/utility of the present approach. Of necessity, we leave such generalizations/extensions to readers with more specialized applications in mind.

Finally, we should comment that while our present idealized analysis explicitly deals with *steady-state* AGT- or rocket-spray combustion, it can yield the expected "sensitivity" of various performance indices to many physical parameters that can be varied at will. While this "parametric sensitivity" information is clearly not *sufficient* for a complete understanding of the dynamics of such a combustor, its value for model-based *control purposes* should not be under-estimated.

Summary Remarks

Of course, in practice, many "trade-offs" must be made in the design of any particular spray combustor, and it is not surprising that such a simple mathematical model will remain explicitly silent about some of them. Yet, as a complementary tool in the hands of a design team, even the simplest form of our idealized spray combustor model can provide useful semi-quantitative if not always fully quantitative results. Moreover, as briefly discussed above, enhancements of the present simple model immediately suggest themselves (e.g., including downstream modules, accounting for heat and droplet loss to the containment walls, adding chemical kinetic sub-models—e.g., to predict "flame-out", and emissions ($CO(g)$, $NO(g)$, $SO_3(g)$, soot,...). In any case, it is hoped that a diverse group of readers will be motivated to extend/adapt these ideas/methods/results in novel ways.

CONCLUSIONS

We have reformulated and extended the now-classical problem of liquid fuel-fed idealized steady-flow combustor performance. New theoretical methods and illustrative numerical results are presented here for the attainable combustion *intensity* and corresponding combustion *efficiency* of well-(jet-) stirred adiabatic vessels fed with a prescribed polydispersed spray. Each droplet in the spray is assumed to evaporate at high pressures according to a *non*-quasi-steady (non-QS) gas-phase energy/mass diffusion-controlled rate law (Rosner and Chang, 1973 and Crespo and Liñán, 1975). As a byproduct, we calculate, by*repeated numerical quadratures alone* , the complete droplet size distribution (DSD) function exiting the chamber, along with its associated dimensionless moments—all potentially of interest for the design of downstream components. We explicitly assume that the volumetric rate of chemical energy release in such "primary" combustion chambers is controlled by the liquid fuel (physical) vaporization process itself (with negligible lags due to propellant droplet HU or vapor-phase ignition). In this instructive asymptotic limit, two decisive non-dimensional parameters are shown to be: (1) a vaporization Damköhler number (i.e., ratio of the mean residence time of the chemically reacting gaseous mixture in the combustion chamber to the reference value of the vaporization lifetime of a droplet with the SMD produced by the fuel injector(s)), and (2) a dimensionless non-QS parameter (proportional to the square-root of the product of phase density ratio and (log of) (1 + fuel volatility ratio)). Even at this level of idealization, our present formulation appears to economically provide much useful information about how spray combustor performance will depend on such inter-active factors as: injector ("atomizer") performance (say, SMD and DSD spread), available chamber volume, V_{ch} (irrespective of shape), inlet (feed) conditions (liquid phase and vapor phase), liquid fuel and oxidizer properties, overall propellant mixture ratio, and chamber pressure. Our results (e.g.,Fig. 2) reveal the existence of an optimum vaporization Damköhler number which maximizes the combustion intensity—with peak intensities being quite sensitive to the non-QS parameter at high pressures. In the process, we have therefore demonstrated the unappreciated consequences of presuming a QS (local d^2) law for individual droplet evaporation rates. Indeed, Fig. 7 clearly shows that, even at modest chamber pressures, the QS-rate law significantly underestimates the achievable combustion efficiency and intensity. Perhaps more attention will have to be paid to this fundamental area in developing more comprehensive spray combustor models for high-intensity applications. This long-overdue example serves to illustrate one of many possible uses of this mathematical model—as an economical "test-bed" for examining the importance of suggested spray combustor approximations and simplifications.

Results from our present model also set instructive bounds to the achievable steady-state performance of "real" spray combustors. With the abovementioned modest enhancements, this approach can be used to complement other, more elaborate design tools, to economically map out the sensitivity of spray combustor performance to a large number of important design or control parameters.

ACKNOWLEDGMENTS

This research was supported in part by NSF via Grant CTS-0522944 at Yale University, along with financial support (for MAZ) from Projects S-0505/ENE/0229 and ENE2005-09190-c04-02 at UNED (Madrid).

REFERENCES

1. Archambault, M.R., Edwards, C.F., McCormack, R.W., 2003. Atomization and Sprays 13, 89.

2. Bartok, W., Heath, C.E., Weiss, M.A., 1960. A.I.Ch.E. Journal 6 (4), 685.

3. Bayvel, L., Orzechowski, Z., 1993. Liquid Atomization. Taylor and Francis, Washington, DC.

4. Birouk, M., Gokalp, I., 2006. Progress in Energy and Combustion Science 32, 408(see, also, 2002. International Journal of Heat and Mass Transfer 45, 37).

5. Bittker, D.A., Brokaw, R.S., 1960. American Rocket Society Journal 30 (2), 179.

6. Boyd, J.P., 2000. Chebyshev and Fourier Spectral Methods. second ed. Dover, New York.

7.   Courtney, W.G., 1960. American Rocket Society Journal 30, 356.

8.   Crespo, A., Liñán, A., 1975. Combustion Science and Technology 11, 9.

9.   Derr, W.S., Mellor, A.M., 1990. Design of Modern Turbine Combustors. Academic Press, London.

10.   Duda, J.L., Vrentas, J.S., 1971. International Journal of Heat and Mass Transfer 14 (3), 395.

11.   Hubbard, G.L., Denny, V.E., Mills, A.F., 1975. International Journal of Heat and Mass Transfer 18, 1003.

12.   Huzel, D.K., Huang, D.H., 1992. AIAA, Progress in Astrodynamics and Aerodynamics Series, vol. 147, 431pp.

13.   Labowsky, M., 1976. Chemical Engineering Science 31, 803. See also, ACS Advances in Chemistry, 1978, Vol 166, pp. 63--79.

14.   Labowsky, M., 1980. Combustion Science and Technology 22, 217.

15.   Law, C.K., 2006. Combustion Physics. Cambridge University Press, New York.

16.   Lefebvre, A.H., 1999. Gas Turbine Combustion. second ed. Taylor and Francis, Washington, DC.

17.   Mattingly, J.D., Heiser, W.H., Pratt, D.T., 2002. Aircraft Engine Design. second ed. AIAA, Reston, VA.

18.   Menon, S., Patel, N., 2006. AIAA Journal 44 (4), 709.

19.   Nuruzzaman, A.S.M., Siddall, R.G., Beer, J.M., 1971. Chemical Engineering Science 26, 1635.

20.   Priem, R.J., Heidmann, M.F., 1960. Propellant vaporization as a design criterion for rocket engine combustion chambers. NASA TR R-67.

21.   Probert, R.P., 1946. Philosophical Magazine 27 (265), 94.

22.   Rosner, D.E., 1972. In: Liquid Propellant Rocket Combustion Instability. NASA SP-194,

23.   p. 74 (Chapter 2.4).

24.   Rosner, D.E., 1989. A.I.Ch.E. Journal 35 (1), 164.

25.   Rosner, D.E., 1997. Chemical Engineering Education (ASEE) 31, 228(see, also, 1998. Chemical Engineering Education (ASEE) 32(1), 82).

26. Rosner, D.E., 2000. Transport Processes in Chemically Reacting Flow Systems. Dover

27. Publications, New York, NY. (fourth printing with 48pp. updates. First edition/printing Butterworth-Heinemann, 1986).

28. Rosner, D.E., 2006. International Journal of Chemical Reactor Engineering 4 (1), Art. No. A21 http://www.bepress.com/ijcre/vol4/A21 [Presented at Tenth International Symposium on Chemical Reactor Engineering, August 2005].

29. Rosner, D.E., Arias-Zugasti, M., 2007. A.I.Ch.E. Journal 53 (7), 1879.

30. Rosner, D.E., Chang, W.S., 1973. Combustion Science and Technology 7 (145--158),

31. Rosner, D.E., Epstein, M., 1970. Journal of Physical Chemistry 74 (22), 4001.

32. Rosner, D.E., Khalil, Y.F., 2000. Journal of Aerosol Science 31 (3), 273.

33. Rosner, D.E., Tassopoulos, M., 1989. A.I.Ch.E. Journal 35 (9), 1497.

34. Rosner, D.E., McGraw, R., Tandon, P., 2003. Industrial and Engineering Chemistry Research 42, 2699.

35. Sangiovanni, J.J., Labowsky, M., 1982. Combustion and Flame 47 (1), 15.

36. Shapiro, A.H., Erickson, A.J., 1956. In: Proceedings of the Heat Transfer and Fluid Mechanics Institute, ASME Transactions, June 21--23, p. 775.

37. Sirignano, W., 1999. Fluid Dynamics of Drops and Sprays. Cambridge University Press, UK.

38. Spalding, D.B., 1959. American Rocket Society Journal 29, 828.

39. Sutton, R.D., Hines, W.S., Combs, L.P., 1972. AIAA Journal 10 (2), 194.

40. Swithenbank, J., Poll, I., Vincent, M.W., Wright, D.D., 1973. Proceedings of the Combustion Institute, 14th Symposium (International) on Combustion, pp. 627--638.

41. Tolpadi, A.K., Aggarwal, S.K., Mongia, H.C., 2000. Numerical Heat Transfer 38 (4), 325. Williams, F.A., 1985. Combustion Theory. second ed. Benjamin Cummings, Menlo Park, CA.

42. Wittig, S., Vohringer, O., Kim, S., 2002. High-Intensity Combustors-Steady Isobaric Combustion. Wiley-VCH and DFG,

Chapter 5

Numerical Estimation of Blowout, Flashback, and Flame Position in MIT Micro Gas-Turbine Chamber

Hamed B. Ganji and Reza Ebrahimi

Combustion and Propulsion Laboratory, Faculty of Aerospace Engineering, K.N. Toosi University of Technology, Daneshgah Blvd., East Vafadar Ave., 4th Sq. of Tehranpars, Tehran 1656983911, Iran

ABSTRACT

Combustion of hydrogen–air mixture has been simulated numerically inside the MIT (Massachusetts Institute of Technology) micro gas-turbine chamber. Blowout, flashback, and flame position have been studied for different equivalence ratios. Some of the considerations in this simulation are applying a 9-species, 19-step hydrogen–air reaction mechanism, thermal coupling of reacting flow and solid structure of the combustor, considering radiation and convection heat loss from the outer surface of the combustor, and exerting physical boundary conditions on 3D geometry of the combustion chamber. To solve the simulating equations for 3D computational fluid dynamics model, finite volume method has been implemented, and parallel processing has been performed on 6 compute nodes. To validate employed simulating models, the simulation results have been compared with experiment results reported from MIT laboratory and also with simulation results obtained by another research team. The comparison shows that using eddy dissipation concept model (EDC) with disabled turbulence productions and turbulent viscosity terms in k and transport equations and solving equations with remaining terms can predict range of mass flow for stable combustion much closer to experimental results (more than 200% improvement in simulation results), which implies that it can be considered as a relatively reliable method for modeling mean reaction rate of micro-combustion.

INTRODUCTION

Recently, many different kinds of micro devices such as micro-spacecrafts, air vehicles, actuators, and robots have got a lot of attention due to their various mission capabilities and also the reduction in energy consumption. Consequently, the needs for micro power-supply are increasing, especially on the power sources with high energy density (Hua et al., 2005a). On the other hand, new technology has made it possible to manufacture and assemble a new generation of micro heat engines for power generation and micro air-vehicle propulsion applications (Spadaccini et al., 2003). The combustion-based micro engines can deliver an output power density tens of times higher than the best lithium batteries ever made (Spadaccini et al., 2003). Some applications for micro-engines can be assumed as micro-propulsion, propulsion system with distributed fine thrust vectors, mobile power generator, micro-refrigerating systems, micro-controller for boundary layer and circulation, and power supply for developing micro-systems. Some of the internal combustion micro-engines and micro-power-supply concepts can be referred to as piston engine, rotary engine (Wankle), gas-turbine engine (Dumand et al., 2005), and thermo photovoltaic system (Chia and Feng, 2007).

The small dimensions of the micro-systems imply that these devices operate in a somewhat unfamiliar parameter regime (Janson et al., 1999). Recently, some experimental and numerical researches have been conducted in different aspects of micro-combustion. 2D transient simulations of methane-fueled micro-reactor with a platinum catalyst wall have been performed by Karagiannidis and Mantzaras (2010). These simulations were performed by varying the inlet pressure, the solid wall thermal conductivity and heat capacity, the inlet velocity, and the equivalence ratio at fuel–lean conditions. They evaluated reaction rates using CHEMKIN code. Some 3D numerical simulations of 3.5 mm wide spiral Swiss roll heat-recirculating combustors have been performed by Chen and Ronney (2011). Their simulation included finite rate chemical reaction of 1-step and 4-step propane–air mixtures, solid-phase conduction and surface-to-surface radiation. They observed that results are surprisingly similar with or without a turbulence model activated (Chen and Ronney, 2011). The combustion of H_2-air in a 2D geometry of a micro-combustor has been simulated numerically by Jejurkar and Mishra, 2010, Jejurkar and Mishra, 2011a and Jejurkar and Mishra, 2011b using the Arrhenius relation for a one-step stoichiometric hydrogen oxidation mechanism. They achieved a self-sustaining combustion under different inlet velocities and wall thermal conductivities without any need for catalyst.

Due to high power density of the micro gas-turbine which can be up to 3000 MW/m^3 (Epstein, 2003), it has been considered as one of the most-favored micro-engine concepts and one of the first power MEMS devices (Maruta, 2011 and Chou et al., 2011). The power density of the MIT micro gas-turbine is approximately 1100 MW/m^3. Spadaccini et al. (2003) obtained temperature contour of the combustion chamber by employing 3D numerical simulation, and applying the chemical kinetics with a 9-species, 20-step hydrogen–air reaction mechanism. Nevertheless, the heat transfer between reacting flow and the solid combustor walls as well as the heat conduction through solid structure was not simulated in their work. Peck (2003) modeled heat transfer through solid structure of the combustor and the heat loss from this structure to the ambient, but the fluid dynamics and the combustion kinetics have not been considered in that modeling. Another numerical simulation of the MIT micro gas-turbine chamber is the work done byHua et al. (2005b). Most of the effective parameters have been included in that simulation such as detailed chemical kinetics, fluid dynamics, heat transfer within solid structure of the combustor, and heat loss to the ambience. However, the predicted inlet mass flow rate for blowout (0.4 g/s) is far from the experiment result (0.12 g/s) reported by Mehra (2000) and also Spadaccini et al. (2003).

The 3D numerical simulation of combustion in the MIT micro gas-turbine chamber presented in this paper involves detailed chemical kinetics, thermal coupling of the reacting flow and solid structure of the combustor, and heat loss to the ambient. To model reaction rate, EDC combustion model has been employed with turbulence generations and turbulent viscosity terms eliminated from the k– viscous model. The effects of eliminating turbulence productions terms while using EDC model have been studied as well as the effects of inlet mass flow rate (velocity) and equivalence ratio on blowout, flashback, flame position, and burnt gas volume. Finally, combustor efficiency graphs have been plotted to be compared with experiment results reported by Mehra (2000) and Spadaccini et al. (2003) and also with simulation results achieved by Hua et al. (2005b).

MODELING EQUATIONS AND APPROACHES

Equations and method used in this research have been described in this section as well as a concise presentation of applied chemical kinetic mechanism and numerical approach. The main assumptions of this simulation are steady-state, no gas radiations, and no surface reaction.

Modeling Equations and Material Properties

Basic modeling equations for a reacting flow are continuity, momentum, energy conservation, species transport, and reaction rate modeling equations. Residence time for GE90 conventional combustor is about 7 ms (Spadaccini et al., 2003); while, due to the miniature volume of the micro-combustor, its residence time is approximately 0.5 ms (Spadaccini et al., 2003 and Epstein, 2003). Consequently, an appropriate reaction model for micro-combustion must be sensitive to this short residence time. Regarding these conditions, EDC model has been chosen for modeling mean reaction rate, which will be discussed more in the next sections.

For all species in this simulation, density has been defined by using the ideal gas model, and specific heat capacity has been determined as a piecewise-polynomial function of temperature. Thermal conductivity, viscosity, and mass diffusivity have been modeled using the kinetic theory (FLUENT 6.3 User's Guide, 2006). Finally, the mixture physical properties have been modeled by applying the mass-weighted-mixing-law. Prior to this study, Norton and Vlachos (2003) have used the mentioned physical property models to simulate the micro scale combustion of premixed methane–air. Thermal conductivity for solid structure of silicon combustor is 149 W/m K. All mentioned modeling equations have been listed in Table 1.

Table 1: Modeling equations and physical properties of materials

Modeling equations	
Continuity	$\nabla \cdot (\rho \vec{v}) = 0$
Momentum	$\nabla \cdot (\rho \vec{v} \vec{v}) = -\nabla p + \nabla \cdot (\equiv \tau) + \rho \vec{g}$
Fluid energy conservation	$\nabla \cdot (\vec{v} \rho h) = \nabla \cdot \left[k_{eff} \nabla T - \sum_j h_j \vec{J}_j \right] + S_h; \quad S_h = -\sum_j \left[\frac{h_i}{M_w} + \int_{T_{ref,i}}^T c_{p,i}\, dT \right] R_i$
Solid (wall) energy conservation	$\nabla \cdot (\kappa^M \nabla L) = 0$
Heat loss equation (to the ambient)	$q = h_{ext}(T_{amb} - T_w) + \epsilon_{ext}\, \sigma\, (T_{amb}^4 - T_w^4)$
Species transport equation	$\nabla \cdot (\rho \vec{v} Y_i) = -\nabla \cdot \vec{J}_i + R_i; \quad \vec{J}_i = -\rho D_{i,m} \nabla Y_i - D_{T,i} \frac{\nabla T}{T}$
EDC model equations	$R_i = \frac{\rho\, \varphi^2}{\tau \left(1 - \varphi^3 \right)} (Y_i^* - Y_i); \quad \xi = 2.1377 \left(\frac{v\varepsilon}{k^2}\right)^{3/4}; \quad \tau = 0.4082 \left(\frac{v}{\varepsilon}\right)^{1/2}$
Arrhenius rates relations	$\sum_{i=1}^N \iota_{i,r} M_i \rightleftharpoons \sum_{i=1}^N \iota_{i,r} M_i; \quad k_{f,r} = A_r T^{\beta_r} e^{-E_r/RT}; \quad k_b = \frac{k_{f,r}}{K_r}$ $k_{e,r} = \Gamma(\upsilon_{j,r} - \upsilon_{j,r}^*) \left[k_{f,r} \prod_{j=1}^N (C_{j,r})^{\eta_{j,r}} - k_{b,r} \prod_{j=1}^N (C_{j,r})^{\eta_{j,r}} \right]; \quad \Gamma = \sum_j^N \upsilon_{j,r} C_j$
RNG k-model transport equations (without turbulence productions and turbulent viscosity terms)	$\frac{\partial}{\partial t} (\rho k u_k) = \frac{\partial}{\partial x_i} \left(\alpha \mu \frac{\partial k}{\partial x_j} \right) - \rho \varepsilon$ $\frac{\partial}{\partial x_i} (\rho \varepsilon u_i) = \frac{\partial}{\partial x_j} \left(\alpha \mu \frac{\partial \varepsilon}{\partial x_j} \right) - 1.68 \rho \frac{\varepsilon^2}{k}$ $\left| \frac{\alpha - 1.3929}{\alpha_0 - 1.3929} \right|^{0.6321} \left| \frac{\alpha + 2.3929}{\alpha_0 + 2.3929} \right|^{0.3679} = 1; \quad \alpha_0 = 1.0$
Physical properties of materials	
Density	$\rho = \dfrac{p_{op}}{\frac{R}{M_w} T}$
Species viscosity	$\mu_i = 2.67 \times 10^{-6} \dfrac{\sqrt{M_w T}}{\sigma_i^2 \Omega_\mu}$
Species thermal conductivities	$k_i = \dfrac{15}{4} \dfrac{R}{M_w} \mu \left(\dfrac{4}{15} \dfrac{c_p M_w}{R} + \dfrac{1}{3} \right)$
Binary mass diffusion coefficient	$D_{ij} = 0.0188 \dfrac{\left[T^3 \left(\frac{1}{M_{w,i}} + \frac{1}{M_{w,j}} \right) \right]^{1/2}}{p_{abs} \sigma_{ij}^2 \Omega_D}$
Mixture physical properties (mass-weighted-mixing-law)	$\mu = \sum_i Y_i \mu_i; \quad k = \sum_i Y_i k_i; \quad c_p = \sum_i Y_i c_{p,i}$

Principles of EDC Model

The EDC model, presents empirical expression for mean reaction rate based on assumption that chemical reaction takes place where the dissipation of turbulent energy occurs (Gran and Magnussen, 1996). These regions, according to energy cascade model, consist of fine structures with characteristic dimensions of the order of Kolmogorov micro-scale in one or two dimensions (Gran and Magnussen, 1996). At size equal to Kolmogorov micro-scale or smaller, no turbulent structure exists due to the fact that in those regions molecular diffusion is faster than turbulence transport (Rehm et al., 2009). Moreover, within those fine structures, reactants are mixed at a molecular scale, therefore, ready for chemical reactions to occur (Magnussen, 1981 and Magnussen, 2005). The fine structure regions thus create the reaction space for non-uniformly distributed reactants (Magnussen, 2005).

EDC model assumes that the fine structures occupy only a fraction of the flow. The reactive volume fraction of the flow is ξ^3, where ξ is expressed as (Gran and Magnussen, 1996)

$$\xi = \left(\frac{3C_{D2}}{4C_{D1}^2}\right)^{1/4} \cdot \left(\frac{\nu\varepsilon}{k^2}\right)^{1/4} \tag{1}$$

Where model constants $_{CD1}=0.134$ and $_{CD2}=0.50$ (Gran and Magnussen, 1996) are obtained by employing energy cascade model, ν is the kinematic viscosity, and k and ε are the turbulent kinetic energy and its dissipation rate respectively. Therefore, EDC model assumes that the flow is divided into a reactive volume fraction (ξ^3), and a non-reactive part or surrounding ($1-\xi^3$). Eq. (1) implies that for low Reynolds regimes, lower turbulent kinetic energy leads to the bigger value of the ξ. The upper limit of the empirical EDC model for reactive portion of the flow is specified as 7.55E–1(Rehm et al., 2009), which has been delineated in Section 4.1.

Chemical reactions take place when reactants are mixed at molecular scale at sufficiently high temperature (Magnussen, 2005). The rate of molecular mixing is determined by the rate of mass transfer between the fine structure regions and the surrounding fluid (Magnussen, 2005). The time scale for mass transfer between fine structures and surroundings is estimated by time scale τ (Gran and Magnussen, 1996)

$$\tau = \left(\frac{C_{D2}}{3}\right)^{1/2} \cdot \left(\frac{\nu}{\varepsilon}\right)^{1/2}$$

(2)

During time τ the fine structures are assumed as perfectly stirred reactor (PSR) under constant pressure and adiabatic conditions (Gran and Magnussen, 1996, Gran et al., 1994, Stefanidis et al., 2006 and Parente et al., 2008) and Arrhenius relations control the reaction progress. To calculate the mass exchange rate between fine structures and surroundings, using relation ξ^3/τ seems to be logical; however, due to the mass exchange between fine structures with adjacent fine structures, Magnussen proposed using ξ^2/τ (Gran and Magnussen, 1996). Therefore, the mean reaction rate can be calculated as

$$R_i = \frac{\rho\xi^2}{\tau}(Y_i^* - Y_i^\circ)$$

(3)

In which, superscript * refers to inside fine structures, and ° denotes the surroundings. Mass fraction for species i inside fine structures, Y_i^*, is calculated by integrating from the Arrhenius relation over time scale . Mean mass fraction for species i is expressed as

$$Y_i = \xi^3 Y_i^* + (1 - \xi^3)Y_i^\circ$$

(4)

Then Eq. (3) can be written based on mean mass fraction as

$$R_i = \frac{\rho\xi^2}{\tau} \cdot \left(\frac{Y_i^* - Y_i}{1 - \xi^3}\right)$$

(5)

Mean reaction rate calculated by Eq. (5) is used as the source term of species transport equation.

Employing EDC Model in Simulating Micro-Combustion

Residence time for micro-combustion chamber is 0.07 of that for GE90 conventional combustor according to Spadaccini et al. (2003) and Epstein (2003). This short residence time of micro-combustor makes it prone to blowout. It is well known that PSR assumption is appropriate to predict the mass flow rate and equivalence ratio of blowout in coupled chemical and thermal systems (Turns, 2000, Chapter 6). EDC model is one of the best available empirical combustion models in which the PSR assumption has been implemented on fine structures where the chemical reactions occur. The empirical-based EDC model puts an upper limit of 0.755 on the reactive volume fraction in a low Re regime flow (Rehm et al., 2009). Within those reactive portions of the flow, reactants are mixed at molecular scale and thus ready for chemical reactions to take place under the control of Arrhenius relations. The required time scale for mass transfer between the non-reactive portion of the flow and the reactive fine structures is calculated based on the flow parameters and experimental constants (Eq. 2). The reactive portions can be considered as perfectly stirred reactor and PSR assumption is employed by the EDC model. EDC mean reaction rate is then calculated by multiplying the mass transfer (between reactive portions of the flow and surrounding) by Arrhenius controlled mass fraction changes inside the reactive volume. EDC mean reaction rate actually restricts the high amount of Arrhenius (chemical) reaction rate by considering the mass flow between non-reactive portion of the flow and the reactive fine structures (Eq. 3).

Molecular scale mixing time is calculated based on EDC empirical constants   and also the flow parameters k and ε. When the complete form of k–ε transport equations are used in this micro-combustion simulating, the value of k and   are calculated to be very large, the time scale value becomes very small, and consequently the EDC-calculated mean reaction rate will be very high, hence combustion will be predicted to occur in the cooling jacket—which is not confirmed by experiment results. It should be considered that the Reynolds number in this problem is approximately 120 in the entrance and it reduces to 20 inside the combustion chamber. This has two effects—first, the viscous losses are high, and second, the flow is laminar (Mehra, 2000). To take account of these low-Reynolds effects through this research, turbulence productions and turbulent viscosity source terms were removed from RNG k–ε transport equations and many cases were simulated. It has been observed that when the turbulence source terms are turned off, regardless of initial values of k and ε at the entrance of the combustor, they dissipate very fast and reach a specific low range. It appears that when the turbulence source terms are omitted from RNG k–ε transport equations, the micro-scale geometry with such a high surface to volume ratio   imposes a significant dissipation on value of k

and ε and causes them to damp to small certain amounts —relatively independent from their initial values. Consequently, what remains from k and ε (which are fairly small amounts) can be considered to be more dependent on micro-scale geometry. As a result of omitting turbulence source terms from k and ε transport equations, the molecular scale mixing time will be calculated higher, the EDC mean reaction rate will be smaller, and the combustion is numerically predicted to take place inside the combustor, which is verified by experiment results presented by Spadaccini et al. (2003) and Mehra (2000).

Therefore, using the empirical-based EDC model with turbulence source terms turned off in k and εtransport equations can be considered as an available engineering approach to apply PSR conditions in the 3D geometry of a micro-combustor and to predict its blowout mass flow rate and equivalence ratio.

The RNG k–ε model has been used in this simulation. While the standard k–ε model is a high-Reynolds-number model, the RNG theory provides an analytically-derived differential formula for effective viscosity that accounts for low-Reynolds-number effects (FLUENT 6.3 User's Guide, 2006). Moreover, non-equilibrium wall function has been employed in this simulation, in which the wall-neighboring cells are assumed to consist of a viscous sub-layer. The viscous sub-layer   prevents the over-estimation of the turbulence effect in vicinity of the combustor walls. After eliminating turbulence productions and turbulent viscosity terms from RNG k and ε transport equations the remaining terms of these equations, which have been presented in Table 1, are k and ε transport by convection, their transport by diffusion, and rate of their destruction.

The independency of solution from initial values of k and ε (while the turbulence source terms are omitted) has been checked by applying different initial values to them. The results have been summarized in Table 2which shows that choosing very different initial values for k and   does not affect the simulation results considerably. However, to avoid inserting unreal turbulence properties in the non-turbulent nature of the micro-combustor, both initial values have been specified equal to 1 for all the simulations performed in this research.

Table 2: Independency of solution from initial values of k and

Initial Value			
k	ε	**Predicted Range Inside Chamber (s)**	**Predicted Max. Temperature (k)**
1	1	1.9E–4 to 3.3E–4	1350
100	100	1.6E–4 to 2.4E–4	1420
1	100	2.1E–4 to 3.5E–4	1370
100	1	1.4E–4 to 2.4E–4	1440

Chemical Kinetic Mechanism

The chemical kinetic mechanism of hydrogen oxidation employed in this simulation has been extracted from the work done by Hua et al. (2005a), which includes 19 reversible elementary reactions and 9 species.Hua et al. (2005a) have validated the employed chemical mechanism by simulating the combustion of premixed hydrogen–air under adiabatic condition and comparing the results with the measurement of adiabatic flame reported in Glassman (1996). Considering the details of production and consumption of free radicals, this mechanism can predict the micro-combustion characteristics very close to the experimental results reported by Mehra (2000) and Spadaccini et al. (2003), which have been shown inSection 4. To evaluate the effect of employed chemical reaction mechanism on the numerical predictions, one other simulation was performed for mass flow rate of 0.05 g/s and equivalence ratio of 0.5 by employing a 33-reaction and 13-species H_2–air chemical reaction mechanism proposed by Jachimowski (1988). Maximum net reaction rate for H_2O was predicted equal to 134 kg/m^3 s by 33-reaction and 13-species mechanism, while it was predicted to be 137 kg/m^3 s by the 19-reaction and 9-species mechanism employed. There is only about 2.2% prediction difference between the two mechanisms. Moreover, maximum temperature was predicted equal to 1350 K by both mechanisms.

Numerical Approach and Solution Convergence

Finite volume method and first-order upwind scheme has been implemented to solve the simulating equations of the 3D computational fluid dynamics model. The modeling equations have been linearized implicitly, and solved using double precision, segregated solver with under-relaxation factors. The segregated solver first solves the momentum equations, then solves the continuity equation, and updates the pressure and mass flow rate afterwards (FLUENT 6.3 User's Guide, 2006). SIMPLE algorithm has been used to deal with the pressure–velocity coupling. The energy and species equations have been solved subsequently, and finally convergence has been checked.

For different simulated cases in this research, the solution convergence has been judged based on scaled residuals criterion (FLUENT 6.3 User's Guide, 2006) which has been set to be less than 1×10^{-6} for momentum and energy equations, and less than 1×10^{-4} for all the rest of modeling equations. About 13 h of parallel processing has been performed on 6 compute nods on a core i7 processor to obtain conclusive converged result for each investigated case.

GEOMETRY, BOUNDARY CONDITIONS, AND MESHING

The real geometry of MIT micro gas-turbine chamber, modeled geometry, specified boundary conditions, and specifications of meshing have been explained in the following subsections.

3D Geometry Modeling

Geometry of 6-wafer MIT micro gas-turbine chamber with slotted inlet has been modeled. Fig. 1 shows a photograph of cross-section of the micro gas-turbine chamber with annular inlet and a schematic view of it is prepared by Epstein et al. (2000). Schematic geometry of slotted-inlet combustor and its major dimensions has been depicted in Fig. 2. Considering the rotational periodic geometry of the combustor, only a sector of it with central angle of 12° has been modeled. Fig. 3 shows the 3D view of the simulated sector of the slotted-inlet combustor.

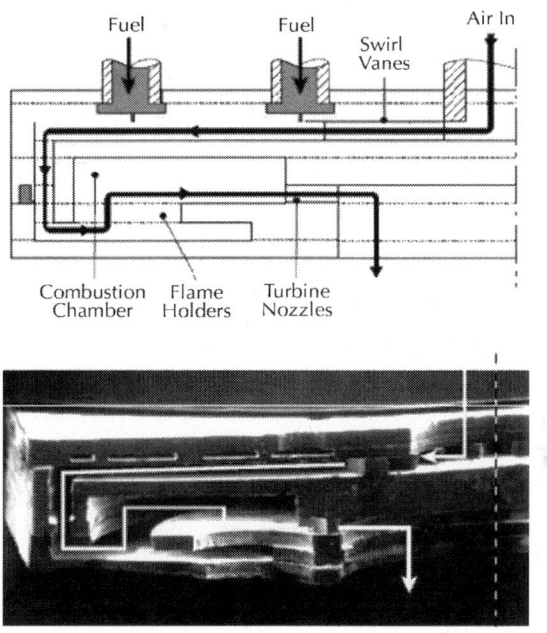

Figure 1: Schematic view and photograph of the MIT 6-wafer micro-combustor cross-section with annular-inlet, extracted from work done byEpstein et al. (2000).

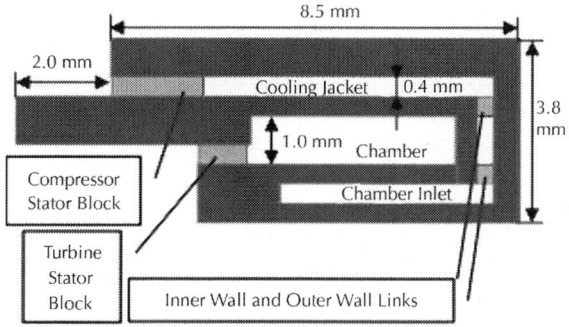

Figure 2: Cross-section of the modeled combustor with slotted-inlet and its major dimensions.

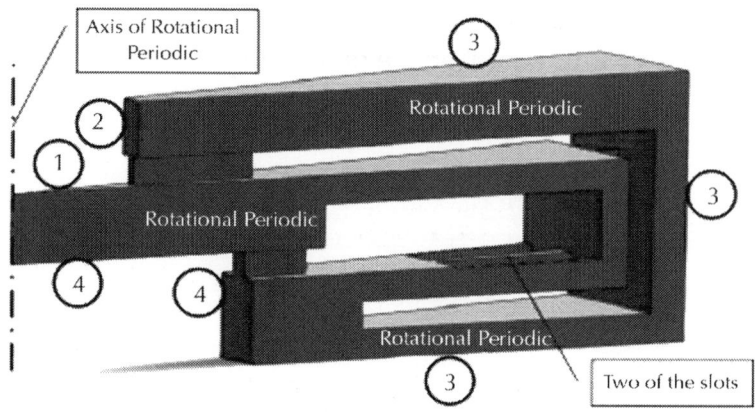

Figure 3: 3D view of solid structure of the combustor; the outer walls are distinguished by numbers. The corresponding thermal convection conditions for each wall have been determined in Table 3.

The MIT micro-combustor (Spadaccini et al., 2003) has been shown in Fig. 1. As shown, after a 90 °e turning, the flow passes through stator blades of the radial compressor and enters the cooling jacket which wraps around the inner wall of the combustion chamber. Passing through the cooling jacket, mixture cools the inner walls, and consequently gets preheated. The mixture then reaches the slotted inlet of the chamber. The modeled inlet consists of 60 slots each having a length of 2.2 mm. The combustion occurs inside the chamber and eventually combustion products (burnt gas) pass through turbine stator blades. In this research the reactive flow has been simulated numerically in the area beginning from just after compressor stator blades to just before turbine stator blades. However, to consider the comprehensive thermal effects of solid structure of the combustor including stator blades, this structure has been modeled as depicted in Fig. 3.

Boundary Conditions

Boundary conditions on the entrance surface of fluid have been set to let the hydrogen–air mixture enter with a specified mass fraction and a defined mass flow rate for each simulated case. MIT test rig has been reported to be an atmospheric combustor without rotor blades (Spadaccini et al., 2003). Inasmuch as simulation results are to be compared with those experiment results, a constant pressure of 101.3 kPa has been imposed on the outlet surface of the fluid. Applied boundary conditions over all the interfaces between fluid and solid structure are no slip, zero-gradient for species, and coupled thermal condition. Due to the rotational periodic characteristic of the modeled sector (solid structure and its contained fluid), therotational periodic boundary condition has been specified on the lateral surfaces of it as shown in Fig. 3. At a periodic boundary, the flow is treated as though the opposing periodic plane is a direct neighbor to the cells adjacent to the first periodic boundary (FLUENT 6.3 User's Guide, 2006). Rotational periodic boundary condition, meanwhile, makes it possible to simulate a swirl in the combustor sector, caused by the exit angle of the compressor stator blades as it has been simulated by Ganji (2010). The axis of rotation is the axis of combustor as shown in Fig. 3.

Regarding external wall numbers demonstrated in Fig. 3, thermal convection conditions are determined inTable 3 which have been extracted from Peck's M.S. thesis (Peck, 2003). In the mentioned thesis, these convection coefficients have been validated for being used in numerical simulating of the heat loss from the solid structure of the MIT micro-combustor. The external wall emissivity (ϵ_{ext}) has been specified equal to 0.85 for silicon walls, as done by (Hua et al., 2005b). Meanwhile, to reach the steady-state solution, initial temperatures of 1200 K and 1000 K were specified for fluid and solid zones respectively.

Table 3: Thermal convective conditions on surfaces of the outer walls, according to Peck (2003)

Wall No.	Specifications of the ambient in the vicinity of the wall	Approximate gas temperature	h (W/m2 K)
(1)	Cool flow, perpendicular to wall	300	772
(2)	Cool flow, tangential to wall	300	98.6
(3)	Natural convection	300	25
(4)	Hot combustion product flow, tangential to wall	Exit temperature ±10%	202.7

Meshing and Partitioning

Approximately 769,600 cells have been used for meshing the fluid and solid structure. A grid with about 1,230,000 computational cells was also tested, which did not change the obtained results. Except for the small wedge-shaped tip of modeled sector, which has been meshed by hexahedral wedge cells, the rest of the whole geometry has been meshed using hexahedral cells. To perform the parallel processing, the whole meshed geometry has been divided into 6 partitions, each processed by a separate compute node on a core i7 Processor.

RESULTS AND DISCUSSION

The presented results are generally divided into four parts: effects of employing EDC model, effects of inlet mass flow, effects of mixture equivalence ratio, and comparison and validation.

Effects of Employing EDC Model and Omitting Turbulence Source Terms from Viscous Model

The reported values of different parameters in this section have been extracted from contours belonging to two different simulations. In both simulations, the inlet velocity is 5 m/s (equal to mass flow rate of 0.050 g/s) with equivalence ratio of 0.5; however, in the first simulation, EDC model has been applied with complete k–ε equations (hereafter called case I), while in the second one, EDC model has been employed with eliminated turbulence productions and turbulent viscosity terms from k–ε equations (hereafter called case II).

For case I , the amount of turbulent kinetic energy, k, around the combustion area is within the range of 29–42 m^2/s^2 and the range for turbulent kinetic energy dissipation rate, ε, is 2.2E+5 to 5.5E+5 m^2/s^3 whereas for case II , the amount of k changes between two low values of 0.28 and 0.34 m^2/s^2, and ε range is 350 to 460 m^2/s^3. Therefore, a very significant drop in amounts of turbulent kinetic energy and its dissipation rate is observable when the turbulence source terms are eliminated.

According to simulation results, EDC time scale, τ, changes between two amounts of 9.3E−6 and 1.3E−5 in combustion chamber for case I . In comparison, τ for case II   is within the range of 1.9E−4 to 3.3E−4 inside the combustion chamber. This increase of τ is expected by regarding Eq. (2) and paying attention to the great decreasing of ε for case II. From the physical point of view, by decreasing turbulent kinetic energy and its dissipation rate, the time needed for molecular scale   mixing is expected to increase. In contrast to τ, which differs for two cases I and II , the amount of EDC volume fraction, ξ, except for a very small area (less than 0.4 mm long) in the entrance of the combustor, is 7.55E−1 for both the cases I and II   all over the fluid. This is attributed to the upper limit of EDC model for ξ value which is specified as 7.55E−1(Rehm et al., 2009).

According to H_2O mean reaction rate contours shown for cases I and II in Fig. 4(a) and (b), maximum reaction rate for case I is 1.04E+3 kg/m³ s; while, for case II, it reduces to 1.37E+2 kg/m³ s. Reduction of predicted reaction rate from1.04E+3 kg/m³ s for case I to 1.37E+2 kg/m³ s for case II can be described by Eq. (5), and considering the fact that the value of ξ is the same for both cases I and II , while time scale τ is bigger for case II. The high amount of reaction rate for case I causes this simulation to predict combustion inside the cooling jacket whereas according to experiment results (Spadaccini et al., 2003), for the same situations, combustion area is limited to inside the combustion chamber. For case II, however, the restricted reaction rate predicts the combustion area to be inside the chamber.

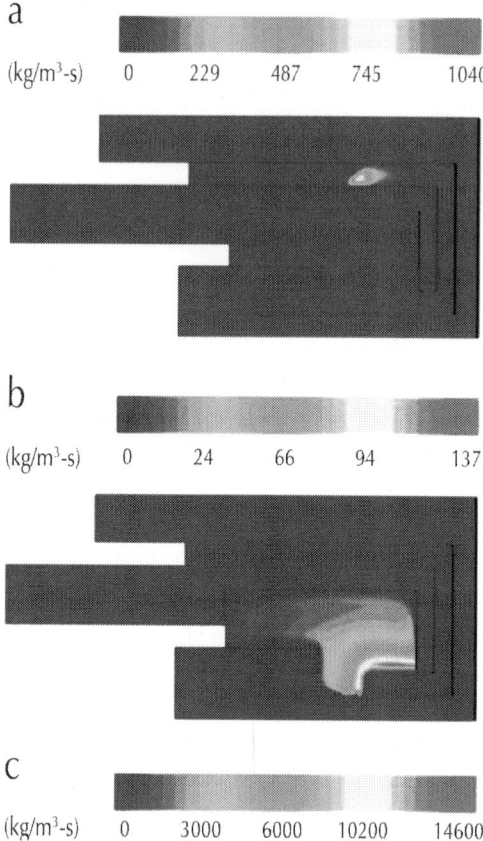

a

(kg/m³-s)   0   229   487   745   1040

b

(kg/m³-s)   0   24   66   94   137

c

(kg/m³-s)   0   3000   6000   10200   14600

Figure 4: Comparison between H$_2$O mean reaction rate contours for: (a) case I, (b) case II, and (c) Arrhenius reaction rate. (a) H$_2$O Mean Reaction Rate contour for case I (EDC with full k–ε), φ = 0.5, flow rate 0.050 g/s, (b) H$_2$O Mean Reaction Rate contour for case II(EDC with turbulence productions and turbulent viscosity turned off in k–ε), φ = 0.5, flow rate 0.050 g/s, (c) H$_2$O Mean Reaction Rate contour predicted by using Arrhenius reaction rate relations, φ = 0.5, flow rate 0.050 g/s.

For further study, another simulation was performed with the same conditions as cases I and II; however, Arrhenius reaction rate relations were employed in this simulation to model the reaction rate. To consider the Soret effect in this resent case, thermal diffusion has also been employed in species transport modeling. It is interesting that in this recent case, due to not restricting the reaction rate by molecular scale mixing time, Arrhenius relations calculate the mean reaction rate as high as 1.46E+4 kg/m^3 s and the combustion occurs just in the beginning of the cooling jacket, which is quite far from the real conditions. Mean reaction rate contour for H_2O has been presented in Fig. 4(c).

It can be concluded that by employing EDC combustion model and simultaneously eliminating turbulence productions and turbulent viscosity terms from RNG k–ε transport equations and solving the remaining terms, the predicted flame zone for micro-combustor is more compatible with real conditions. This can be attributed to calculate a more realistic reaction rate by taking account of molecular scale mixing time, τ, and making PSR assumption while preventing inserting unreal turbulence in laminar micro-reacting flow.

Effects of Mass Flow Rate

To study the effects of mass flow rate, the equivalence ratio of hydrogen–air mixture has been set to be 0.5, and the micro-combustion has been simulated by imposing different inlet velocities corresponding to different inlet mass flow rates. Fig. 5 shows maximum temperature, exit temperature, and wall temperaturediagrams for 13 different mass flow rates. Maximum temperature in this paper refers to the highest gas temperature observable in the temperature contours, exit temperature is the area-weighted average of temperature over the outlet surface of the combustor simulated sector, and wall temperature is the area-weighted average of temperature over the external face of the combustor walls. It should be mentioned that wall temperature is relatively uniform due to the fine dimensions of combustor, and also because of the high thermal conductivity of silicon. More details about what happens in the diagrams shown in Fig. 5 are described by Fig. 6 which displays temperature contours for equivalence ratio of 0.5. There are different experimental methods for determining flame surface such as observing luminous part of the flame, shadowgraph picture, and Schlieren picture. Schlieren surface lies quite early in the flame is more readily definable than most images, and is recommended and preferred by many workers (Glassman, 1987). In this numerical simulation however, as the mean reaction rate contour is available, the surface on which the value of mean reaction rate is maximum has been considered as flame surface. An approximation of this surface position has been shown by dotted line as an estimation of flame surface on all presented temperature contours. Based on the inlet mass flow rate magnitude, there are four different conditions observable for combustion, burnt gas zone, and flame position as follows:

- Mass flow rate from 0.017 to 0.040 g/s: in this range, flame surface and burnt gas is placed downstream of the slots, as shown in Fig. 6(a). For inlet mass flow rate lower than 0.017 g/s, the simulation predicts a thermal quench of combustion which can be attributed to low rate of heat generation and high heat loss to the surroundings. In other words, small volume of burnt gas zone and its high surface to volume ratio describes the reason for this thermal quench.

- Mass flow rate from 0.040 to 0.080 g/s: according to Fig. 6(b), increasing the inlet velocity within this range expands the volume occupied by burnt gas. In order to burn more amounts of hydrogen–air mixture in unit time which is equivalent to increasing the inlet power, the burnt gas volume and temperature increase. This results in a higher wall temperature as well.

- Mass flow rate from 0.080 to 0.114 g/s: increasing the inlet mass flow rate within this range will push combustion area a little forward into the chamber as shown in Fig. 6(c). For inlet mass flow rate bigger than 0.080 g/s, the limitations of reaction rate and flame speed restrict the flame ability to burn the fuel–air mixture with such a high inlet velocity, and flame extends its surface by forming a concave curve to gain the capability of burning higher amounts of mass flow rate which have been shown by dotted line in Fig. 6(c). Burning of the increased amount of inlet fuel–air mixture inside the contracted volume increases the maximum temperature. The increased temperature, according to the Arrhenius relations, and causes the increase of reaction rates. The final result is a self-sustaining combustion with a smaller burnt gas volume and higher temperature pushed a little forward into the combustion chamber. The contracted burnt gas volume on the other hand, lessens the adjacency of burnt gas and combustor wall, and wall temperature reduces a little, which can be seen in Fig. 5.

- Mass flow rate more than 0.114 g/s: increasing the inlet mass flow rate just a little more than 0.114 g/s will lead to an immediate flame blowout. This sudden blowout is detectable by the steep slope of the end part of diagrams shown in Fig. 5. Temperature contour for inlet mass flow rate of 0.114 g/s is presented in Fig. 6(d), which illustrates the simulation result just before complete disappearing of the flame. After some more iteration, this small hot spot will be completely removed. This blowout can be explained by considering the fact that increasing inlet mass flow rate—which is equal to increasing the inlet velocity—decreases the residence time; consequently, the required time for molecular scale mixing   (time scale τ) is no longer available and reactants leave the chamber without being burnt. It should be added that by increasing the inlet mass flow rate more than 0.114 g/s the prior explained concave curve of the flame expands and causes the effective volume of the chamber to reduce, which decreases the residence time even more.

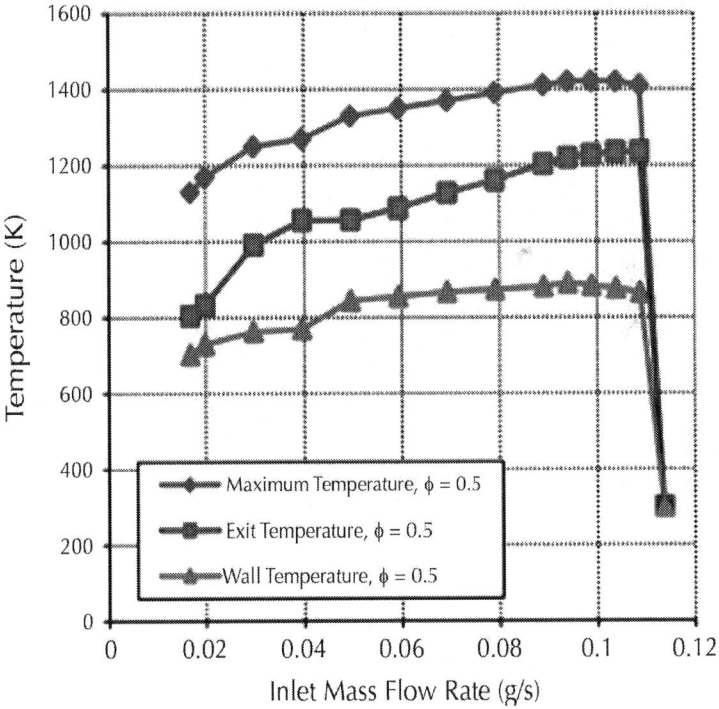

Figure 5: Maximum temperature, exit temperature, and wall temperature diagrams over the numerically predicted range of mixture mass flow for stable combustion, for equivalence ratio of 0.5.

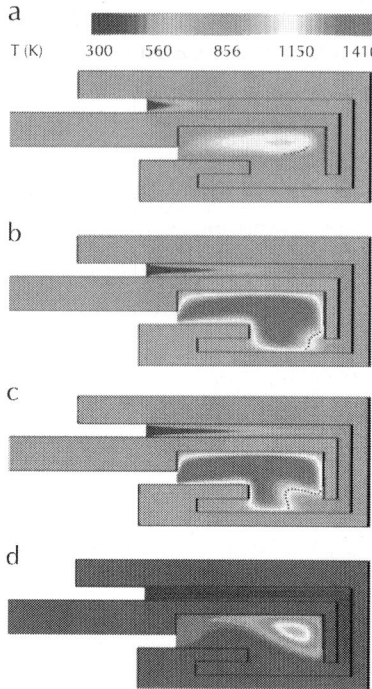

Figure 6: Temperature contours for different inlet mass flow rates with equivalence ratio of 0.5. Dotted lines show approximately the flame surface on which the mean reaction rate has its maximum value. (a) Inlet mass flow rate: 0.017 g/s; Maximum temp.: 1130 K; Wall temp.: 702 K; Exit temp.: 805 K; φ = 0.5, (b) Inlet mass flow rate: 0.080 g/s; Maximum temp.: 1390 K; Wall temp.: 873 K; Exit temp.: 1160 K; φ = 0.5, (c) Inlet mass flow rate: 0.109 g/s; Maximum temp.: 1410 K; Wall temp.: 865 K; Exit temp.: 1240 K; φ = 0.5, (d) Inlet mass flow rate: 0.114 g/s; the flame quenches completely after some more iterations; φ = 0.5.

Effects of Equivalence Ratio

One important reason for burning a fuel–lean mixture in micro-combustor is to avoid high flame temperature which can damage the silicon structure of the micro-combustor. In order to study the effect of fuel/air ratio on micro-combustion characteristics, flame surface, and burnt gas zone, the micro-combustion has been simulated for two more equivalence ratios of φ=0.45 and φ=0.6.

Equivalence Ratio of 0.45

Stable combustion was simulated for a fuel–lean mixture with equivalence ratio of 0.45, and for ratio of 0.4 the combustion was not self-sustaining. Fig. 7 illustrates the temperature diagrams predicted by simulation for equivalence ratios of 0.45 and 0.6. A comparison between results achieved for φ=0.45 (Fig. 7) with results obtained for φ=0.5 (Fig. 5) shows firstly that the range of mixture mass flow for stable combustion is more restricted for φ=0.45, and secondly, the overall maximum temperature for φ=0.45 is lower than that for φ=0.5. These results are not against expectations as it is known that by lowering the equivalence ratio below the stoichiometric conditions, both flame speed and maximum temperature are reduced.

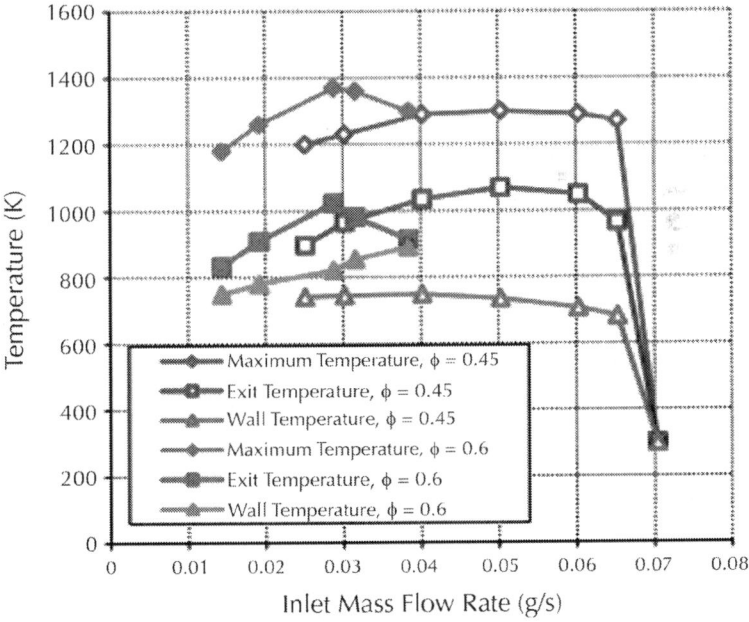

Figure 7: Maximum temperature, exit temperature, and wall temperature diagrams over the numerically predicted range of mixture mass flow for stable combustion, for equivalence ratios of 0.45 and 0.6.

Two temperature contours for equivalence ratio of 0.45 are illustrated in Fig. 8 which shows that by increasing the inlet mass flow rate from 0.025 to 0.050 g/s, the volume of burnt gas, flame surface (shown by dotted line), and maximum temperature increase. However, according to Fig. 7, increasing the inlet mass flow rate above 0.050 g/s reduces both maximum temperature and volume of burnt gas. Increasing the inlet mass flow rate is equivalent to decreasing residence time, which leads to incomplete chemical reaction and lowers the maximum temperature. Finally, at inlet mass flow rate of 0.070 g/s, the sudden flame blowout occurs.

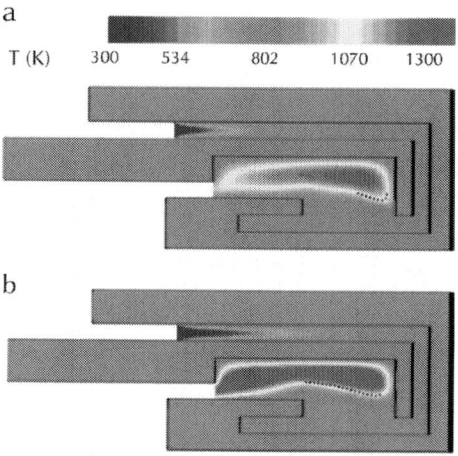

Figure 8: Temperature contours for different inlet mass flow rates with equivalence ratio of 0.45. Dotted lines show approximately the flame surface on which the mean reaction rate has its maximum value. (a) Inlet mass flow rate: 0.025 g/s; Maximum temp.: 1200 K; Wall temp.: 741 K; Exit temp.: 897 K; φ = 0.45, (b) Inlet mass flow rate: 0.050 g/s; Maximum temp.: 1300 K; Wall temp.: 735 K; Exit temp.: 1070 K; φ = 0.45.

Equivalence Ratio of 0.6

Another series of simulations was carried out by applying the entrance mixture with equivalence ratio of 0.6. According to these results, the overall predicted temperature for φ=0.6 is higher than the predicted temperatures for two previous equivalence ratios of 0.5 and 0.45 as can be seen in Fig. 5 and Fig. 7. For equivalence ratio of 0.6, as shown in Fig. 9(a), under relatively low mixture mass flow of 0.032 g/s, the combustion takes place downstream of the cooling jacket and a relatively bulky volume of hot burnt gas appears. Fig. 9(b) shows that by increasing mixture mass flow from 0.032 g/s to 0.038 g/s, wall temperature rises from 854 K to 892 K; as a result, hydrogen–air mixture is preheated as much as getting ignited before entering the combustion chamber and flame flashback draws the flame back into the cooling jacket. When flashback takes place, a great amount of the generated heat is transferred to the jacket wall due to its low width, which not only imposes unallowable temperature to the silicon wall, but also increases heat loss to the surroundings. In addition, as the combustion zone approaches the cooling jacket, it moves further away from the outlet of the combustor, and the combustor outlet temperature decreases, which consequently lessens the combustor efficiency. For equivalence ratio of 0.6, Fig. 7 depicts that when flashback occurs, at mass flow rate of 0.038 g/s, the wall temperature increases, while the maximum temperature and the exit temperature decrease.

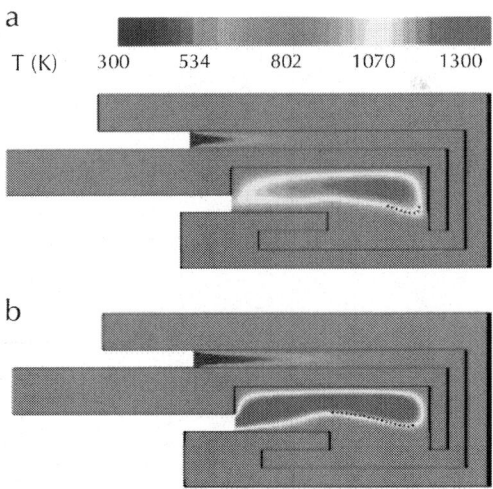

Figure 9: Temperature contours for different inlet mass flow rates with equivalence ratio of 0.6. Dotted lines show approximately the flame surface on which the mean reaction rate has its maximum value. (a) Inlet mass flow rate: 0.032 g/s; Maximum temp.: 1360 K; Wall temp.:

854 K; Exit temp.: 982 K; φ = 0.6, (b) Inlet mass flow rate: 0.038 g/s; Maximum temp.: 1300 K; Wall temp.: 892 K; Exit temp.: 914 K; φ = 0.63.

According to Fig. 5, for equivalence ratio of 0.5 and inlet mass flow rate of 0.094 g/s, the wall temperature was as high as 890 K close to flashback wall temperature for φ=0.6, but flashback has not been predicted for equivalence ratio of 0.5. It must be noticed that for φ=0.5, the high inlet velocity (9.5 m/s) prevents the combustion zone from propagating inside cooling jacket—in other words, the mixture does not have enough time to get preheated and ignited inside the cooling jacket. Moreover, for lower equivalence ratios (less than 1), flame speed is more restricted (Kuo, 1986), which makes the flame flashback less probable.

Comparison and Validation

To compare the obtained simulation results with experiment results reported by Mehra (2000) and also bySpadaccini et al. (2003) and with the results achieved by Hua et al. (2005b), comparative graphs of combustor efficiency have been shown in Fig. 10 for equivalence ratio of 0.5. Overall combustor efficiency can be calculated as below (Spadaccini et al., 2003)

$$\eta_{comb} = [(\dot{m}_a + \dot{m}_f)h_2 - \dot{m}_a h_1]/\dot{m}_f h_f$$

(6)

Where, subscript "1" refers to the combustor inlet and subscript "2" indicates the combustor outlet. Fig. 10 clarifies that the simulation performed in this study has predicted the range of mixture mass flow for stable combustion much closer to experiment results. Differences between simulation and experiment results are 7% and 8% for blowout and thermal quenching, while for Hua et al. (2005b) the differences are 240% and 70% respectively. The other important characteristic of the simulated micro-combustor is the fast blowout of flame at specific mass flow rate, which has been predicted in this study by steep slope of end part of the graph. Experiment results reported by MIT laboratory also admit that the flame blowout actually occurs very fast. However, the graph plotted by Hua et al. (2005b) ends with a mild slope, which does not imply such an abrupt blowout. The improvements of simulation results obtained in this research can most likely be attributed to taking account of molecular scale mixing time in reaction rate modeling.

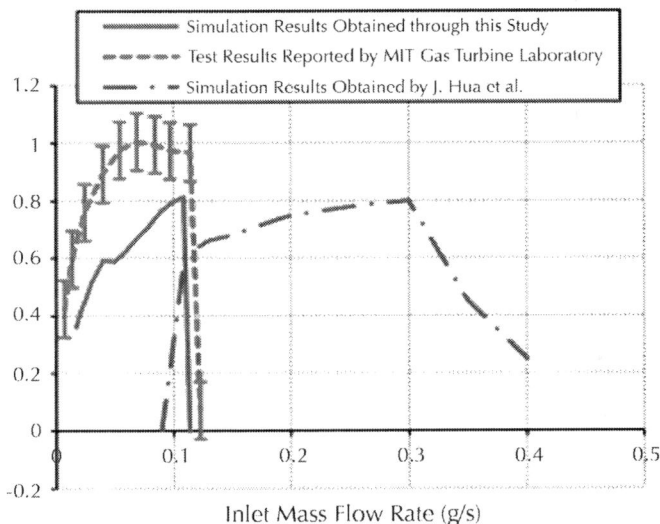

Figure 10: Micro-combustor efficiency graphs—a comparison between simulation results obtained through this study, Hua et al. (2005b)simulation results, and experiment results including 8.9% uncertainty (Mehra, 2000 and Spadaccini et al., 2003) for equivalence ratio of 0.5.

Fig. 11 compares the combustor efficiency graphs which have been plotted based on the simulation results with experimental graphs (Mehra, 2000 and Spadaccini et al., 2003) for different equivalence ratios. This comparison clarifies some points as follows:

- The predicted range of mixture mass flow for stable combustion is comparable with experiment results for different equivalence ratios. For equivalence ratio of 0.45 predicted range is 0.025–0.070 g/s, while experiment results admit the range of 0.006 to 0.086 g/s, for φ=0.5 the prediction is 0.017–0.114 g/s and experiment result is 0.007–0.123 g/s, and for φ=0.6 predicted range is 0.014–0.038 g/s and experiment shows the range of 0.006–0.049 g/s. Minimum difference between predictions and experiment results is 7% which belongs to blowout in φ=0.5, and maximum difference belongs to flashback prediction at mass flow of 0.038 which is 23%.

- Flame flashback has been numerically predicted to occur only for φ=0.6 and not for equivalence ratios of 0.45 and 0.5, which is confirmed by experiment results.
- Efficiencies calculated based on the simulation results are a little different from those reported by MIT laboratory. For example, for equivalence ratio of 0.5, when the mass flow rate is 0.114 g/s, the difference is approximately 15%, and for mass flow rate of 0.050 g/s, the difference is about 30%.

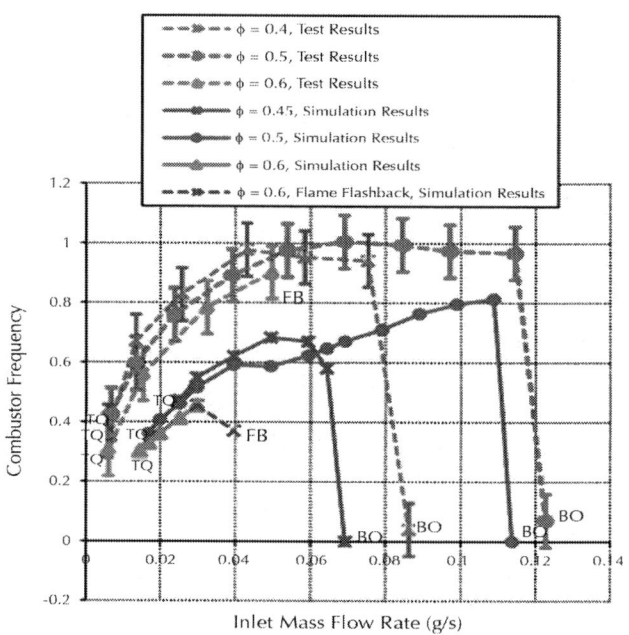

Figure 11: Micro-combustor efficiency graphs for different equivalence ratios —the simulation results obtained through this study are compared with experiment results including 8.9% uncertainty (Mehra, 2000 and Spadaccini et al., 2003). Abbreviations on the graph are TQ: Thermal Quench, BO: Blowout, and FB: Flashback.

According to Mehra (2000 p. 174), there is 8.9% uncertainty in the efficiency diagrams plotted based on experiment results, which has been shown in Fig. 10 and Fig. 11. Peck (2003) also reports that there is ±130 K uncertainty in exit gas temperature diagrams obtained through the experiments. This uncertainty is attributed to difficulties in temperature measuring in micro-scales. Regarding the considerable heat loss from micro-combustor, due to its large surface to volume ratio, and considering the fact that efficiency cannot be more than 100%, it can be concluded that the numerical under-prediction of efficiency is not as much as it appears in the comparison diagrams shown in Fig. 10 and Fig. 11. On the other hand, not taking account of wall effects in applied chemical kinetic mechanism and also convection coefficients which have been used may affect the numerically predicted combustor efficiency. All mentioned points explain probable causes of the observed differences between simulation and experiment results.

CONCLUSIONS

EDC model has some special properties which seem appropriate for modeling reaction rate in a micro-combustor and predicting its blowout and flashback. These properties can be referred to as considering the molecular scale mixing time, applying PSR assumption, and dealing with Kolmogorov micro-scale . In order to estimate a value for to be used by EDC model, transport equations of k– viscous model should be solved. Considering the low Reynolds number of micro-flow, turbulence productions and turbulent viscosity source terms have been eliminated from RNG k– transport equations in this research. Then it has been observed that, due to miniature size of micro-combustor and its high surface to volume ratio, the micro-scale geometry of the combustor causes the amounts of k and to dissipate to very small certain amounts which are relatively independent of their initial values. EDC model then can calculate a restricted reaction rate by taking account of the molecular scale mixing time and applying PSR assumption, which can predict blowout, flashback, and flame position close to reality for different mass flow rates and equivalence ratios.

The effects of mass flow rate have been studied throughout the numerically predicted range of mass flow for stable combustion. It has been observed that temperature rises by increasing the inlet mass flow rate, and further increasing of mass flow will ultimately result in a sudden flame blowout. Equivalence ratio effects have also been studied by comparing results for equivalence ratios of 0.45, 0.5, and 0.6 to each other. It was observed that the maximum range of mass flow for stable combustion belongs to equivalence ratio of 0.5 and increasing or decreasing the equivalence ratio from 0.5 reduces this range, which is in compliance with experiment results reported by Mehra (2000); Spadaccini et al. (2003). Moreover, it has been demonstrated that the flame flashback takes place only in equivalence ratio of 0.6, which is confirmed by the same experiment results.

Numerical simulation results must be quantitatively reliable in order to be used as a part of design process. The agreement between this simulation and experiment results implies that EDC combustion model with disabled turbulence productions and turbulent viscosity source terms in RNG k– transport equations can be considered as an available engineering approach which can help to apply PSR conditions and to model the steady-state mean reaction rate of micro-combustion for predicting blowout, flashback, flame position, and range of mass flow for stable micro-combustion in different equivalence ratios. Regarding the high surface to volume ratio and consequently high amount of heat loss from micro-combustor, flame stabilization might be effected by the transient process of ignition, which can be a subject of future study.

ACKNOWLEDGMENTS

The Authors are grateful to Professor Mani Fathali for his kind cooperation in delineating the detailed facts about viscous models.

REFERENCES

1. Chen, C.H., Ronney, P.D., 2011. Three-dimensional effects in counterflow heatrecirculating combustors. In: Proceedings of the Combustion Institute 33, pp. 3285–3291.

2. Chia, L.C., Feng, B., 2007. The development of a micropower (micro-thermophotovoltaic) device. J. Power Sour. 165, 455–480.

3. Chou, S.K., Yang, W.M., Chua, K.J., Li, J., Zhang, K.L., 2011. Development of micro power generators—a review. Appl. Energy 88, 1–16.

4. Dumand, C., Guidez, J., Orain, M., Sabel'nikov, V.A., 2005. Specific problems of micro-turbine for micro-drones application. In: European Conference for Aerospace Sciences (EUCASS) Session 5.8: Aero Engines: Turbojets.

5.   Epstein, A.H., Jacobson, S.A., Protz, J.M., Frechette, L.G., 2000. Shirtbutton-sized gas turbines: the engineering challenges of micro high speed rotating machinery. In: 8th International Symposium on Transport Phenomena and Dynamics of Rotating Machinery (ISROMAC-8), Honolulu, March.

6.   Epstein, A.H., 2003. Millimeter-scale, mems gas turbine engines. In: Proceedings of ASME Turbo Expo Power for Land, Sea, and Air, June 16–19, 2003 Atlanta, Georgia, USA.

7.   FLUENT 6.3 User's Guide , 2006 Fluent Inc., Lebanon, New Hampshire, USA.

8.   Ganji, H.B., 2010. Numerical Simulation of Combustion in a Micro Gas-Turbine Chamber (M.S. thesis). KN Toosi University of Technology, Tehran.

9.   Glassman, I., 1987. Combustion. Academic Press Inc., London, pp. 128–129. (LTD).

10.   Glassman, I., 1996. Combustion. Academic Press, California.

11.   Gran, I.R., Melaaen, M.C., Magnussen, B.F., 1994. Numerical simulation of local extinction effects in turbulent combustor flows of methane and air. In: 25th Symposium (International) on Combustion/the Combustion Institute 25, pp. 1283–1291.

12.   Gran, I.R., Magnussen, B.F., 1996. A numerical study of a bluff-body stabilized diffusion flame. Part 2 Influence of combustion modeling and finite-rate chemistry. Combust. Sci. Technol. 119, 191–217.

13.   Hua, J., Wu, M., Kumar, K., 2005a. Numerical simulation of the combustion of hydrogen–air mixture in micro-scaled chambers. Part I: fundamental study. Chem. Eng. Sci. 60, 3497–3506.

14.   Hua, J., Wu, M., Kumar, K., 2005b. Numerical simulation of the combustion of hydrogen–air mixture in micro-scaled chambers Part II: CFD analysis for a micro-combustor. Chem. Eng. Sci. 60, 3507–3515.

15.   Jachimowski, C.J., 1988. An analytical study of the hydrogen–air reaction mechanism with application to scramjet combustion. NASA Tech. Pap., 2791.

16.   Janson, S.W., Helvajian, H., Breuer, K., 1999. MEMS, Microengineering and aerospace systems. In: AIAA the American Institute of Aeronautics and Astronautics, Inc. Paper No. 99-3802.

17. Jejurkar, S.Y., Mishra, D.P., 2010. Numerical characterization of a premixed flame based annular microcombustor. Int. J. Hydrogen. Energy. 35, 9755–9766.

18. Jejurkar, S.Y., Mishra, D.P., 2011a. Thermal performance characteristics of a microcombustor for heating and propulsion. Appl. Therm. Eng. 31, 521–527.

19. Jejurkar, S.Y., Mishra, D.P., 2011b. Flame stability studies in a hydrogen–air premixed flame annular microcombustor. Int. J. Hydrogen Energy. 36, 7326–7338.

20. Karagiannidis, S., Mantzaras, J., 2010. Numerical investigation on the start-up of methane-fueled catalytic microreactors. Combust. Flame 157, 1400–1413.

21. Kuo, K.K., 1986. Principles of Combustion. John Wiley & Sons, Inc., pp. 329–330. Magnussen, B.F., 1981. On the structure of turbulence and a generalized eddy dissipation concept for chemical reaction in turbulent flow. In: 19th American Institute of Aeronautics and Astronautics Aerospace Science Meeting, Missouri, USA, January 12–15.

22. Magnussen, B.F., 2005. The eddy dissipation concept a bridge between science and technology. In: ECCOMAS Thematic Conference on Computational Combustion, Lisbon, June 21–24.

23. Maruta, K., 2011. Micro and mesoscale combustion. In: Proceedings of the Combustion Institute 33, pp. 125–150.

24. Mehra, A., 2000. Development of High Power Density Combustion System for a Silicon Micro Gas Turbine Engine (Ph. D. thesis), Massachusetts Institute of Technology, USA, pp. 40, 101, 102, 174.

25. Norton, D.G., Vlachos, D.G., 2003. Combustion characteristics and flame stability at the microscale: a CFD study of premixed methane/air mixtures. Chem. Eng. Sci. 58, 4871–4882.

26. Parente, A., Galletti, C., Tognotti, L., 2008. Effect of the combustion model and kinetic mechanism on the MILD combustion in an industrial burner fed with hydrogen enriched fuels. Int. J. Hydrogen. Energy 33, 7553–7564.

27. Peck, J., 2003. Development of a Catalytic Combustion System for the MIT Micro Gas Turbine Engine (M.S. thesis), Massachusetts Institute of Technology, USA, pp. 55–56.

28. Rehm, M., Seifert, P., Meyer, B., 2009. Theoretical and numerical investigation on the EDC-model for turbulence–chemistry interaction at gasification conditions. Comput. Chem. Eng. 33, 402–407.

29. Spadaccini, C.M., Mehra, A., Lee, J., Zhang, X., Lukachko, S., Waitz, I.A., 2003. High power density silicon combustion systems for micro gas turbine engines. J. Eng. Gas Turbines Power 125, 709–719.

30. Stefanidis, G.D., Merci, B., Heynderickx, G.J., Marin, G.B., 2006. CFD simulations of steam cracking furnaces using detailed combustion mechanisms. Comput. Chem. Eng. 30, 635–649.

31. Turns, S.R., 2000. An Introduction to Combustion. McGraw-Hill, pp. 189–195.

Sand Production Control by Chemical Consolidation

M.R. Talaghat, F. Esmaeilzadeh, and D. Mowla

Chemical and Petroleum Engineering Department, Shiraz University, Shiraz, Iran

ABSTRACT

Production of sand during oil production causes severe operational problem for oil producers. Several techniques have been used for sand production control in sandstone reservoirs. These techniques are divided into four groups including; standard rig operation with retrievable packer; tubing-conveyed string; coiled tubing and long zone/selective treatment. Several consolidating materials, such as, crude oil coke and nickel plating, have been used in the past by researchers. At present, the chemical binders, such as; phenol resin, phenol–formaldehyde, epoxy, and furan or phenol–furfural provide cementation. The main object of this paper is to present a suitable resin to be used as consolidating agent in the Asmari oil wells of the Ahwaz and Mansoori oil fields. Considering the locally produced resins and required desirable characteristics of resins for consolidation process, six types of resins including; two types of epoxy resins, three types of phenol–formaldehyde resins and a single type of acrylic resin were selected for testing. Different core samples were made by mixing these resins and their hardening agents with a sand sample provided from the Ahwaz and Mansoori oil fields in various percentages. The core samples are tested for permeability, porosity and compressive strength measurement. The obtained experimental data showed that only for a given type of phenol–formaldehyde resins, the permeability and porosity of the core samples are retained in acceptable values and their compressive strength become greater than 3000 psi.

INTRODUCTION

About 70% of the total world's hydrocarbons are located in poorly consolidated reservoirs (Nouri et al., 2003a and Nouri et al., 2003b). These rocks are usually relatively young in geologic age, and are unconsolidated because natural processes have not cemented the rock grains together by mineral deposition (Dees, 1993). As a result, many reservoirs are susceptible to sand production. This is particularly significant in cases that involve sufficient changes of in-situ stresses, high oil production rates, collapse of hole cavities, and presence of water in the formation (Morita et al., 1987, Wang et al., 1991, Kooijman et al., 1996 and Abass et al., 2002). Field observations reported in the literature indicate volumetric concentration of sand in oil pipe systems varies from 1% to 40% (Morita and Boyd, 1991, Geilikman et al., 1994, Tronvoll et al., 2001, Abass et al., 2002, Nouri et al., 2003a and Nouri et al., 2003b). Sand production imposes high costs and many nuisances on the oil industries. Sand production might lead to the erosion of down-hole and surface equipment, thus creating severe safety problems including loss of well control, blowouts, fires and production shut-in (Tippie and Kohlhass, 1973). Sand production might also restrict the quantity of the oil withdrawn from the reservoir leading to the need for a higher number of wells to achieve the overall hydrocarbon recovery, and consequently to additional cost (Moricca et al., 1994). To produce oil and gas from poorly consolidated reservoirs, it is necessary to employ "sand control" methods in the wells. The methods for controlling the migration of sand from unconsolidated formations are some times divided into mechanical and chemical classifications (Dees, 1992 and Hugh and Ramos, 1995). Various sand consolidation methods have been employed to prevent or inhibit sand movement with the fluids produced from hydrocarbon-bearing earth formations. Packing the formation with resin-coated particulated solids, wetting the unconsolidated sand with a bonding resin, and placing resin-treated sand between the loose sand in the formation and the well bore to form a screen are chemical methods. The methods have met with varying degrees of success. A dispersion sand consolidation mixture is one in which a consolidating fluid consists of a hydrocarbon carrier, a resin or a resin-forming mixture dispersed in it together with a quantity of particulated solids (Dees et al., 1992, Dees, 1993 and Appah, 2003). The resin consolidation processes have been classified in various ways. Minimum preparation time at well site, low injection pressure, short cure time before restoring well to production, high compressive strength of resulting matrix, good resistance to deterioration from well fluids and commonly used treating fluids and high retained permeability are desirable characteristics for a consolidation process (Hugh and Ramos, 1995). Several types of resins are presently used in the sand control art. Examples of hardenable organic resins which are suitable for use in accordance with this subject are epoxy resins, polyester resins, phenol–formaldehyde resins, urea–formaldehyde resins, furan resins, urethane resins and mixtures of such resins (Sain, 1962, Dees et al., 1992, Dees, 1993, Jennings et al., 1994 and Shu, 1994; Todd et al., 2001, Appah, 2003, Nguyen, 2004 and Nguyen, 2006). Polymerization of resins is caused by catalysts or curing agents. Sand consolidation with resins has been practiced for many years. Resins are forced into the formations by high pressures instantaneously applied when perforations are formed in the casing of wells or when pressures are released from tubing in wells (Weaver and Murphey, 1990; Dees, 1992 and Dees et al., 1992). Chemical methods have several important advantages over mechanical methods, but the high cost of the resins and the difficulties in obtaining sufficiently uniform injection of chemicals have limited application to relatively short intervals of perforations (Dees, 1993). The hardenable resin on the deposited particulate solids caused or permitted to harden whereby a consolidated permeable particulate solid pack is formed between the well bore and loose or incompetent sand in the formation. Among the commercially available processes for consolidating incompetent formations some are developed by service companies and some developed by the research affiliates of companies, whose primary business is the production, refining and marketing of oil. For example "Sanset process" is a development of Esso Production Research Co. It is also known as BCP (Base Catalyzed Process) and as "Humble Process". Phenol–formaldehyde resin is used in this process. It may be applied to formations with temperatures between 29.5 °C and 94 °C. It may be applied at higher temperatures under some circumstances. The consolidated formation may have a compressive strength of 200 atm. and retain 50% of the original permeability. The pumping time or placement time is controlled by the quantity of curing agent added when the resin is mixed. The chemical reaction is exothermic and the resin constituents must be refrigerated immediately before and during mixing. For lower temperatures, several days may be necessary. An enriched formulation is available for greater strength (Hugh and Ramos, 1995). Other commercial available processes for consolidation formations are indicated in Table 1 (Sparlin, 1976, Graham et al., 1982, Dees et al., 1992, Hugh and Ramos, 1995, Nguyen and Dusterhoft, 1999, Dewprashad et al., 2000, Danican et al., 2006 and Nguyen, 2006).

Table 1: Different commercial available consolidation processes

Company	Process	Base of applied resin	Range of temperatures	Range of compressive strength	Permeability to original (%)
			°F	(PSI)	
ESSO	Sanset	Phenol–formaldehyde	85–200	> 3000	50
Shell	Eposand	Epoxy	100–220	> 5000	67
Chevron	–	Epoxy	50–250	> 7000	50
Dow	K-series	Phenol–formaldehyde	280	> 3000	70
Dow	Sandlock IV	Epoxy	Max. 175	–	–
Halliburton	Sanfix	Furan	40–400	> 3000	90
Halliburton	CONPAC	Furan	80 to > 300	> 3000	70
Halliburton	CONPAC II	Furan	40–400	> 3000	70
Continental oil	Sanchek	Furan–phenolic	–	3000	–

In this work, the performance of six types of resin upon the consolidation of sand stone samples produced from Ahwaz and Mansoori oil fields is investigated by measuring the permeability, porosity and the compressive strength.

MATERIALS

Considering the locally produced resins and required desirable characteristics of resins for consolidation process, two types of epoxy resins, novolak phenol–formaldehyde resin, resole phenol–formaldehyde resin, modified phenol–formaldehyde resin and an acrylic resin were selected for testing. The sand sample utilized was provided from the Ahwaz and Mansoori oil fields. Some characteristics of this sand sample are given in Table 2.

Table 2: Some characteristics of Ahwaz and Mansoori sand sample

Characteristic	Unit
Permeability (mD)	500
Porosity (%)	15
Cohesive strength (Mpa)	3.74
Friction angle (degree)	30
Young modulus (GPA)	5
Poison ratio	0.48
Components	Quartz, dolomite, calcite, clay
Particle size (μm)	90–1000

PROCEDURE

A mold suitable for core making in cylindrical form was constructed by PVC tubes and plates. The sand samples are mixed with different percentage of a given resin and its appropriate hardening agent in the mold. After curing, the mold is cut and the obtained core is polished and sized for measurement of different properties. This procedure was repeated for different types of resins with different percentages (10–90%) of sand. Each core sample was tested for three different properties as follows.

Permeability and Porosity Measurement

An experimental apparatus was constructed for measuring the permeability of core samples. Each core sample with 10 cm length and 2.75 cm diameter is placed in the apparatus and for each run the required parameters such as inlet and outlet pressure, volume flow rate of water and cross sectional area of the sample were recorded. To prevent the radial flow through the core, it was coated with RTV adhesive and Teflon band. The absolute permeability was calculated using Darcy's equation in the form of

$$K=(Q \cdot \mu \cdot L)/A\Delta p \tag{1}$$

In this work, the porosity of the core samples was determined by gravitational and helium method.

Compressive Strength Measurement

Compressive strength of the core samples was measured by Instron press apparatus using the ASTM E9-89a procedure. According to this procedure, the samples are compressed until the failure (crushing or fracturing) is occurred. As a result, the compressive strength is determined by dividing the maximum stress upon the cross sectional area at or before failure. The core samples with a length/diameter ratio of 1 were used in this work for determination of the compressive strength.

RESULTS AND DISCUSSION

Hydrocarbons are often found in unconsolidated rocks. These rocks are usually relatively young in geologic age, and are unconsolidated because natural processes have not cemented the rock grains together by mineral deposition. To produce oil and gas from these reservoirs, it is necessary to employ "sand control" methods in the wells. Mechanical and chemical methods are two types of the sand control methods. Chemical methods employ a liquid resin which is injected from a well-bore into the unconsolidated rock surrounding the well. The resin is catalytically polymerized to form a porous, permeable rock mass. With most resins, it is necessary to inject a displacement fluid, which is a fluid following the resin which is not miscible with the resin. Chemical methods have several important advantages over mechanical methods, but the high cost of the resins and the difficulties in obtaining sufficiently uniform injection of chemicals have limited application to relatively short intervals of perforations.

The main object of this work is to present a suitable resin to be used as a consolidating agent in the Asmari oil wells of the Ahwaz and Mansoori oil fields. For this object, the performance of six types of resin upon the consolidation of sand stone samples is investigated by measuring the permeability, porosity and the compressive strength. For this object, Considering the locally produced resins and required desirable characteristics of resins for consolidation process, two types of epoxy resins, novolak phenol–formaldehyde resin, resole phenol–formaldehyde resin, modified phenol–formaldehyde resin and an acrylic resin were selected for experiments. The sand sample utilized was provided from the Ahwaz and Mansoori oil fields that some characteristics of this sand sample are given in Table 2.

For core making, a mold suitable in cylindrical form was constructed by PVC tubes and plates. The sand samples are mixed with different percentage (10 to 90%) of a given resin (including acrylic resin, epoxy resin, novolak phenol–formaldehyde resin, resole phenol–formaldehyde resin and modified phenol–formaldehyde resin) and its appropriate hardening agent in the mold. After curing, the mold is cut and the obtained core is polished and sized for the measurement of different properties. This procedure was repeated for different types of resins with different percentages (10–90%) of the sand. For selection of desired resin, two core samples containing 0 and 50 wt.% sand stones are made with six types of the resins namely two types of epoxy resins, novolak phenol–formaldehyde resin, resole phenol–formaldehyde resin, modified phenol–formaldehyde resin and acrylic resin. The initial results of the core samples show that the acrylic resin and the novolak phenol–formaldehyde resin are not suitable for consolidating the sand stones. Since these resins are cured by air or at a temperature of 150 °C, respectively. Also, these resins do not have any curing agent. Then, the effects of four other types of resin with different percentage of sand stones upon the performance of consolidation by measuring the permeability, porosity and compressive strength are investigated. An experimental apparatus was constructed for measuring the permeability of core samples. Each core sample with 10 cm length and 2.75 cm diameter is placed in the apparatus and for each run the required parameters such as inlet and outlet pressure, volume flow rate of water and cross sectional area of the sample were recorded. To prevent the radial flow through the core, it was coated with RTV adhesive and Teflon band. The absolute permeability was calculated using Eq. (1). For more accuracy in permeability measurement of core samples, the experiment for each core sample was repeated at least 3 tests. The results of permeability measurement for pure resins including epoxy resin type A, epoxy resin type B, phenol–formaldehyde resin type C(Resole type), and modified phenol–formaldehyde resin is shown in Fig. 1. As can be seen in this figure, the absolute permeability of the core sample made by the pure modified phenol–formaldehyde resin is higher than those made by the other types of pure resins. Also, the results of permeability measurement for core samples made by epoxy resin type A, epoxy resin type B, phenol–formaldehyde resin type C, and modified phenol–formaldehyde resin with different percentages (0–90%) of the sand stone was provided from the Ahwaz and Mansoori oil fields are shown in Fig. 2, Fig. 3, Fig. 4 and Fig. 5, respectively. According to these figures, the absolute permeability of the core samples made by the modified phenol–formaldehyde resin is higher than those made by the other types of resins. Also, the permeability decreases as the sand% in core samples for different resins increases. The obtained experimental data show that the curing of samples made by the modified phenol–formaldehyde resin is along with increasing of the temperature and volume, simultaneously. Moreover, comparison between the measured permeability for the core samples made by epoxy resin type A, epoxy resin type B, phenol–formaldehyde resin type C, and modified phenol–formaldehyde resin with different percentages (0–90%) of the sand stone is shown in Fig. 6. According to this figure, the absolute permeability of the core samples made by the modified phenol–formaldehyde resin is higher than those made by the other types of resins for same percentage of sand. Also, as can be seen, the absolute permeability value of the core samples including more than 40% sand made by epoxy resin type A, epoxy resin type B and phenol–formaldehyde resin type C are close to zero. The permeability of core samples made by the modified phenol–formaldehyde resin is between 1500 and 3500 mD and this value is retained in acceptable values since based on data obtained from Iranian Oil Company, the average absolute permeability for the Ahwaz and Mansoori oil fields is about 2000 mD, hence, porosity% of samples was measured for this resin, only. The porosity of the core samples was determined by gravitational and helium method. The effect of sand% variation upon the porosity measurements is shown in Fig. 7. According to this figure, the porosity% of the core samples made by the modified phenol–formaldehyde resin is within 38%–65%. As shown in Fig. 7, the porosity% obtained by two different methods of gravitational and helium are approximately near or equal to each other.

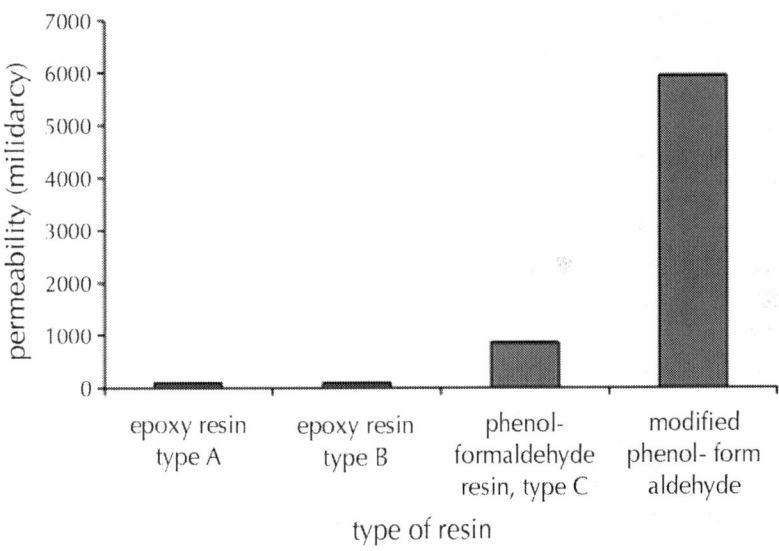

Figure 1: Comparison of permeability for different types of pure resins.

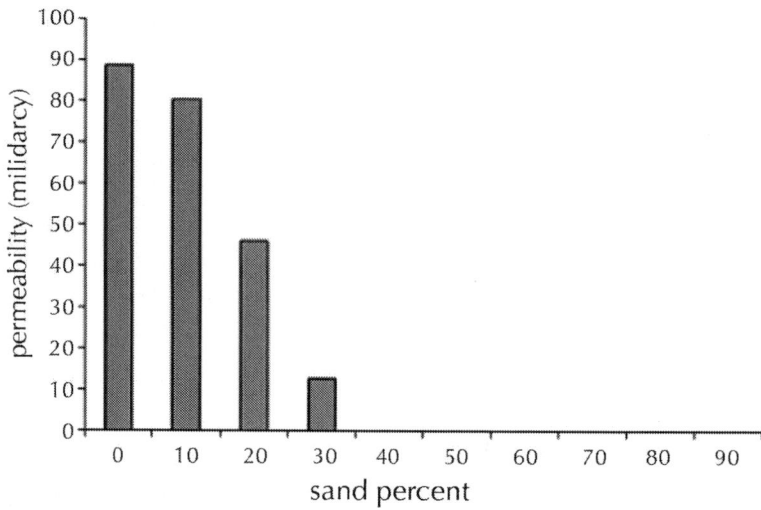

Figure 2: Variations of permeability with sand% in the samples (for epoxy resin: type A).

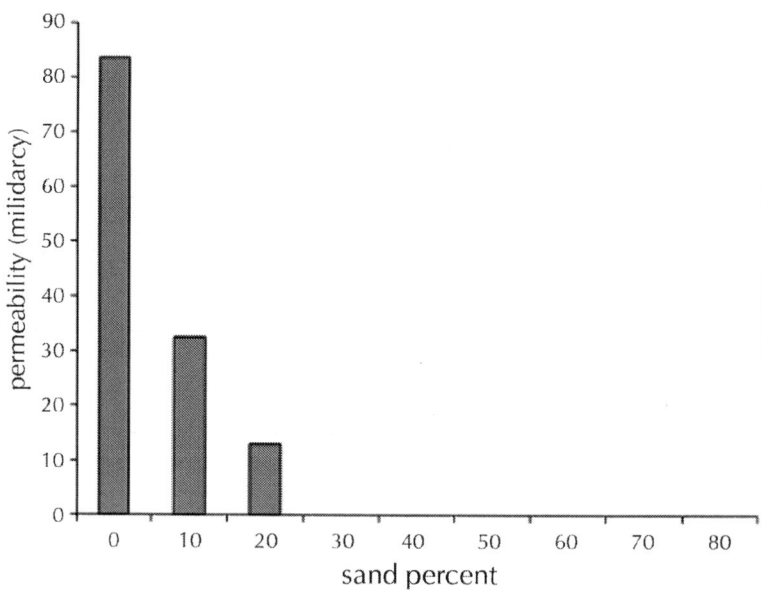

Figure 3: Variations of permeability with sand% in the samples (for epoxy resin: type B).

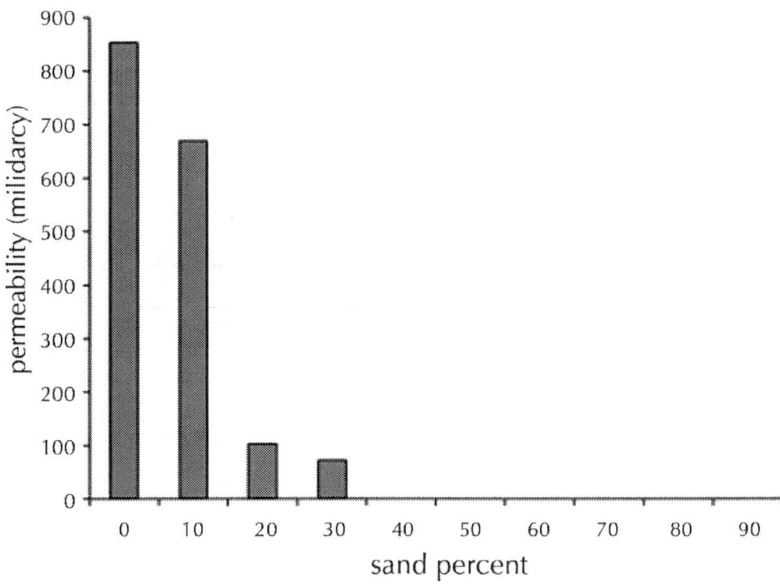

Figure 4: Variations of permeability with sand% in the samples (for phenol–formaldehyde resin: type C).

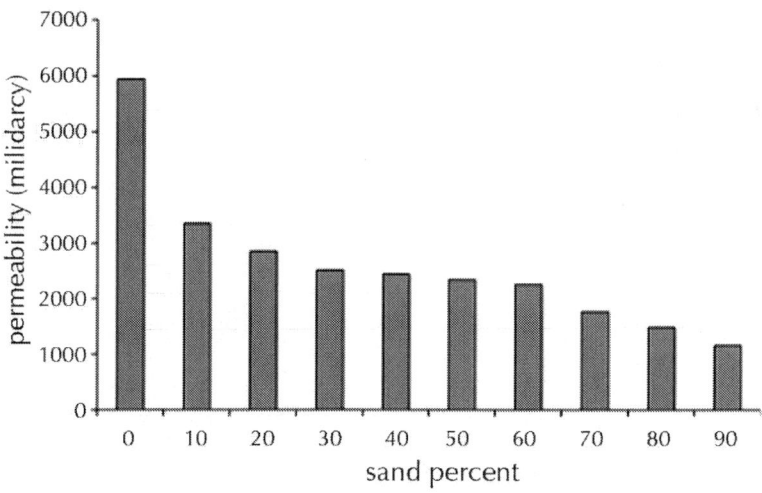

Figure 5: Variations of permeability with sand% in the samples (for modified phenol–formaldehyde).

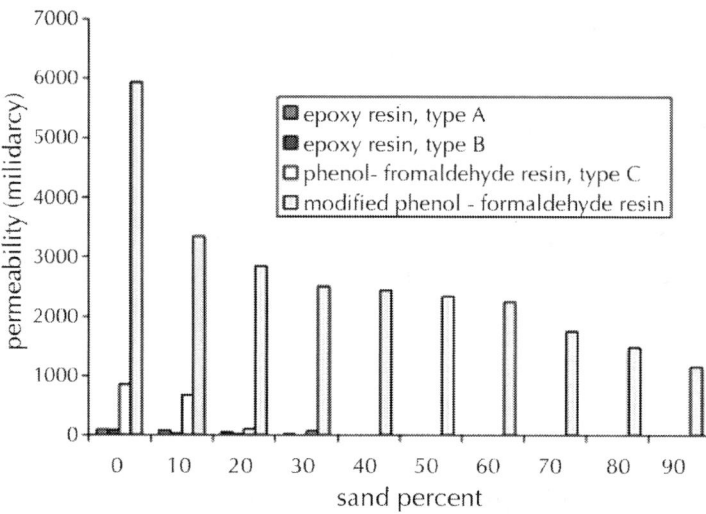

Figure 6: Comparison of permeability for different samples.

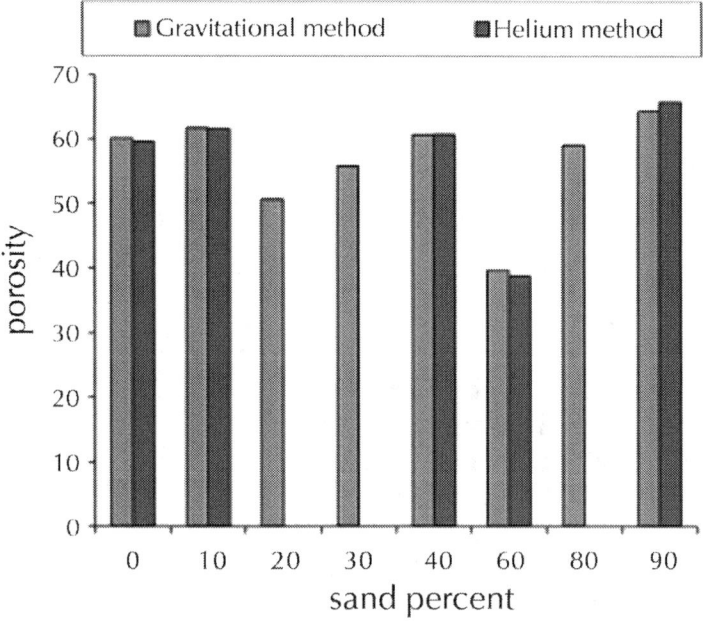

Figure 7: Variations of porosity with sand% in the samples (for modified phenol–formaldehyde resin).

Compressive strength of the core samples was measured by Instron press apparatus using the ASTM E9-89a procedure. According to this procedure, the samples are compressed until the failure (crushing or fracturing) is occurred. The core samples with a length/diameter ratio of 1 were used in this test for determination of the compressive strength. The results of compressive strength measurement for the core samples made by epoxy resin type A, epoxy resin type B, phenol–formaldehyde resin type C, and modified phenol–formaldehyde resin is shown in Fig. 8.

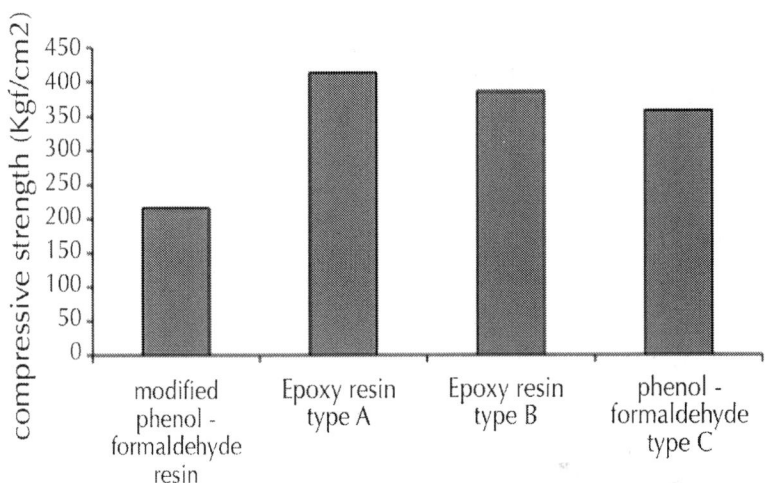

Figure 8: Variations of compressive strength for different pure resins.

As can be seen, the compressive strength of the all core samples is more than 225 Kgf/cm^2. Also, the compressive strength of the core sample made by the modified phenol–formaldehyde resin is higher than those made by the other types of pure resins. The effect of sand% variation upon the compressive strength measurements for core samples made by the modified phenol–formaldehyde resin is shown in Fig. 9. According to this figure, the compressive strength of the core samples made by the modified phenol–formaldehyde resin is greater than 220 Kgf/cm^2 (3000 psi)and the compressive strength of the core samples including 40 or 50 sand% is higher than the other percentages. Therefore, the results of experimental data shown that the modified phenol–formaldehyde resin with the porosity percentage between

38% and 65%, the permeability between 1500 and 3500 mD, and the compressive strength greater than 3000 psi, is a suitable resin for consolidating.

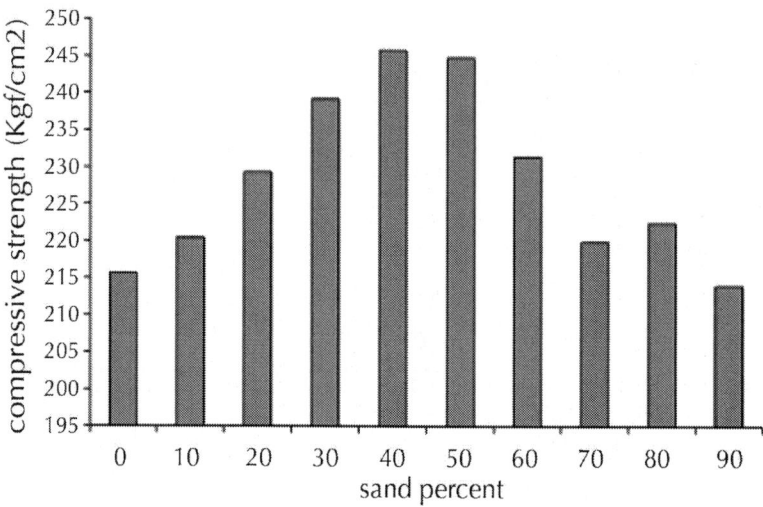

Figure 9: Variations of compressive strength with sand% in the samples (for modified phenol–formaldehyde resin).

CONCLUSIONS

The results of experimental data show that the permeability of the core samples made by modified phenol–formaldehyde resin is between 1500 and 3500 mD, their porosity is between 38% and 68% and their compressive strength is greater than 3000 psi. Therefore, the modified phenol–formaldehyde resin is suitable for consolidating of the Ahwaz and Mansoori oil field formations. Although it seems that this resin could be used for consolidation of other types of sands, but experiments should be conducted for each types of sands stone.

REFERENCES

1.   Abass, H.H., Nasr-El-Din, H.A., BaTaweel, M.H., 2002. Sand Control: Sand Characterization, Failure Mechanisms, And

Completion Methods, SPE Paper 77686 Presented at the SPE Annual Technical Conference And Exhibition San Antonio, Texas.

2.   Appah, D., 2003. Sand Consolidation Experience in Niger Delta. IE (I) Journal-CH, vol. 84,pp. 1–4.

3.   Danican, S., Salamat, G., Pena, A., Nelson, E., 2006. Method of Completing Poorly Consolidated Formations. United States Patent 7013973.

4.   Dees, J.M., 1992. Sand Control in the Wells with Gas Generator And Resin. United States Patent 5101900.

5.   Dees, J.M., 1993. Method of Sand Consolidation with Resin. United States Patent 5178218.

6.   Dees, J.M., Begnaud, W.J., Sahar, N.L., 1992. Sand Control with Resin and Explosive. United States Patent 5145013.

7.   Dewprashad, B.T., Nguyen, P.D., Schreiner, K.L., 2000. Compositions And Methods For Consolidating Unconsolidated Subterranean Zones. United States Patent 6016870.

8.   Geilikman, M.B., Dusseault, M.B., Dullien, F.A., 1994. Sand Production As A Viscoplastic Granular Flow. SPE Paper 27343 Presented at the International Symposium on Formation Damage Control Lafayette, Louisiana.

9.   Graham, J.W., Sinclar, A.R., Brandt, J.L., 1982. Method of Treating Well Using ResinCoated Particles. United States Patent 4336842.

10. Hugh, J.A., Ramos, J., 1995. Guidelines To Sand Control. Report, Halliburton Services from Halliburton Co.

11. Jennings, A.R., Sprunt, E.S., Timmer, R.S., 1994. Method of Sand Consolidation. United States Patent 5363,917.

12. Kooijman, A.P., Van den Hoek, P.J., Bree, Ph. De., Kenter, C.J., Zheng, B.V.Z., Khodaverdian, M., 1996. Horizontal Wellbore Stability And Sand Production In Weakly Consolidated Sandstones. SPE Paper 36419 Presented at the SPE Annual Technical Conference and Exhibition Denver, Colorado. Fig. 8. Variations of compressive strength for different pure resins. Fig. 9. Variations of compressive strength with sand% in the samples (for modified phenol–formaldehyde resin).

13. M.R. Talaghat et al. / Journal of Petroleum Science and Engineering 67 (2009) 34–40 39

14. Moricca, G., Ripa, G., Sanfilippo, F., Santarelli, F.J., 1994. Basin Scale Rock Mechanics: Field Observations Of Sand Production. SPE Paper 28066 Presented at the SPE/ISRM Rock Mechanics in Petroleum Engineering Conference Delft, The Netherlands.

15. Morita, N., Boyd, P.A., 1991. Typical Sand Production Problems: Case Studies And Strategies For Sand Control. SPE Paper 22739 Presented at the 66th Annual Technical Conference and Exhibition of the Society of Petroleum Engineers Dallas, Texas.

16. Morita, N., Whitfill, D.L., Fedde, Ø.P., Løvik, T.H., 1987. Parametric Study Of Sand Production Prediction, Analytical Approach. SPE Paper 16990 Presented at the 62nd Annual Technical Conference and Exhibition of the Society of Petroleum Engineers Dallas, Texas.

17. Nguyen, P.D., 2004. Methods Of Completing Wells In Unconsolidated Formations. United States Patent 6776236.

18. Nguyen, P.D., 2006. Methods For Forming A Permeable And Stable Mass In A Subterranean Formation. United States Patent 6997259.

19. Nguyen, P.D., Dusterhoft, R.G., 1999. Methods Of Completing Wells In Unconsolidated Subterranean Zones. United States Patent 6003600.

20. Nouri, A., Vaziri, H., Belhaj, H., Islam, R., 2003a. Effect of Volumetric Failure On Sand Production In Oil-Wellbore. SPE Paper 84448 Presented at the SPE Asia Pacific Oil and Gas Conference and Exhibition Jakarta, Indonesia.

21. Nouri, A., Vaziri, H., Belhaj, H., Islam, R., 2003b. Comprehensive Transient Modeling Of Sand Production In Horizontal Wellbores, SPE Paper 84500 Presented at the SPE Annual Technical Conference and Exhibition Denver, Colorado.

22. Shu, P., 1994. Consolidation Agent And Method. United States Patent 5362318.

23. Sain, H.H., 1962. Sand Consolidation With A Base-Catalyzed Plastic. API Paper No. 926-7-F, Spring Meeting, Southern District, Houston. Tex.

24. Sparlin, D.D., 1976. A New Sand Control Technique For Old Sand Problems. API paper No. 906-12-H, Spring Meeting, Southwestern District.

25. Tippie, D.B., Kohlhass, C.A., 1973. Effect Of Flow Rate On Stability Of Unconsolidated Producing Sands, SPE Paper 4533 Presented at the 48th Annual Fall Meeting of the Society of Petroleum Engineers of AIME Las Vegas, Nevada.

26. Todd, B., Slabaugh, B.F., Powell, R.J., Yaritz, J.G., 2001. Resin Composition And Methods Of Onsolidating Particulate Solids In Wells With Or Without Closure Pressure. United States Patent 6311773.

27. Tronvoll, J., Dusseault, M.B., Sanfilippo, F., Santarelli, F.J., 2001. The Tools Of Sand Management, SPE Paper 71673 Presented at the 2001 SPE Annual Technical Conference and Exhibition New Orleans, Louisiana.

28. Wang, Z., Peden, J.M., Damasena, E.S.H., 1991. The Prediction Of Operating Conditions To Constrain Sand Production From A Gas Well. SPE Paper 21681 Presented at the Production Operations Symposium Oklahoma City.

29. Weaver, J.D., Murphey, J.R., 1990. Method of particulate consolidation. United states Patent 4936385.

Chapter 7

A Review of an Expert System Design for Crude Oil Distillation Column Using the Neural Networks Model and Process Optimization and Control Using Genetic Algorithm Framework

Lekan Taofeek Popoola[1], Gutti Babagana[2], and Alfred Akpoveta Susu[3]

[1]Department of Petroleum and Chemical Engineering, Afe Babalola University, Ado-Ekiti, Nigeria

[2]Department of Chemical Engineering, University of Maiduguri, Maiduguri, Nigeria

[3]Department of Chemical Engineering, University of Lagos, Lagos, Nigeria

ABSTRACT

This paper presents a comprehensive review of various traditional systems of crude oil distillation column design, modeling, and simulation, optimization and control methods. Artificial neural network (ANN), fuzzy logic (FL) and genetic algorithm (GA) framework were chosen as the best methodologies for design, optimization and control of crude oil distillation column. It was discovered that many past researchers used rigorous simulations which led to convergence problems that were time consuming. The use of dynamic mathematical models was also challenging as these models were also time dependent. The proposed methodologies use back-propagation algorithm to replace the convergence problem using error minimal method.

INTRODUCTION

Crude oil distillation is the separation of hydrocarbons in crude oil into fractions based on their boiling points which lie within a specified range [1]. The separation is done in a large tower that is operated at atmospheric pressure. The tower contains a number of trays where hydrocarbon gases and liquids interact. The liquids flow down the tower and the gases up [2]. The fractions that rise highest in the column before condensing are called light fractions, and those that condense on the lowest trays are called heavy fractions [3]. Figure 1 gives a schematic representation of the process overview of a crude oil distillation column. This crude oil distillation column can be designed using artificial neural network. An artificial neural network (ANN) is an information processing paradigm that is inspired by the way biological nervous systems (such as the brain) process information. It is composed of a large number of highly interconnected processing elements (neurons) working in unison to solve specific problems [4] (Figure 2).

Artificial neural networks initially grew from the full understanding of some ideas and aspects about how biological systems work, especially the human brain. In biology, the cell body of neuron is called the soma. The spine-like extensions of the cell body are dendrites. They usually branch repeatedly and form a bushy tree around the cell body and provide connections to receive incomeing signals from other neurons.

The axon extends away from the cell body to provide a pathway for outgoing signals. Signals are transferred from one neuron to another through a contact point called a synapse [3]. Like biological neuron networks, ANN is made of neuron and synaptic connections which are highly simplified abstracts of their counterparts in real neural networks (Figure 3). The neurons in the ANN are usually called units, node, or processing elements, and the efficacy of the synaptic connection, which is a measurement of excitability and inhabitability, are usually called weights. Neural networks systems (NNS) are typically organized in layers. Layers are made up of a number of interconnected nodes (artificial neurons or processing element), which contains an activation function. The data are presented to the networks via the input layer, which communicates to one or more hidden layers where the actual processing is done through a system of weighted connections. The hidden layers are then linked to an output layer, which generates the output [5, 6].

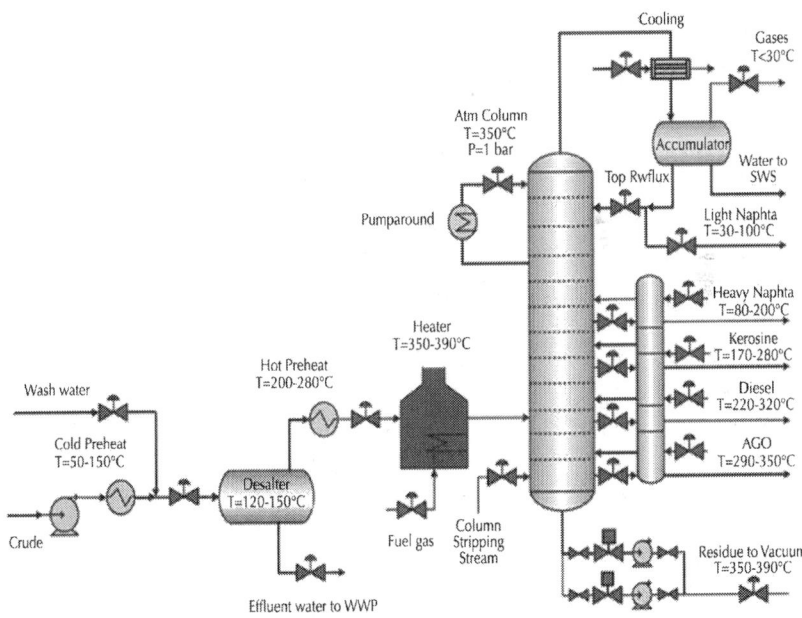

Figure 1: Process overview of crude oil distillation column.

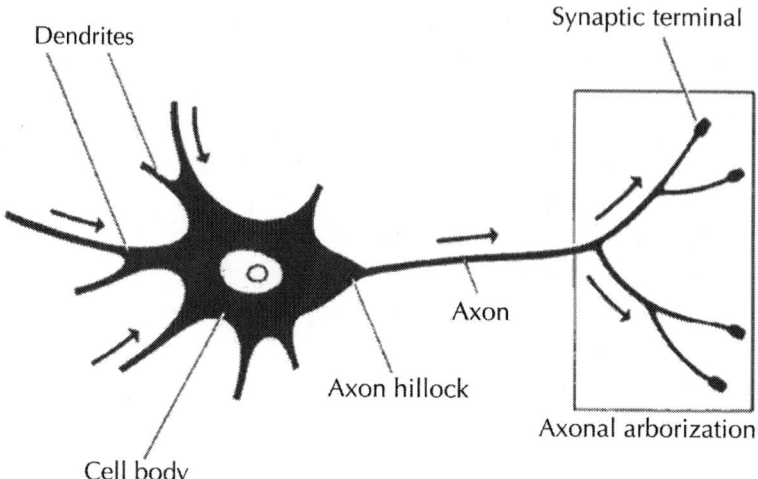

Figure 2: Basic neuron.

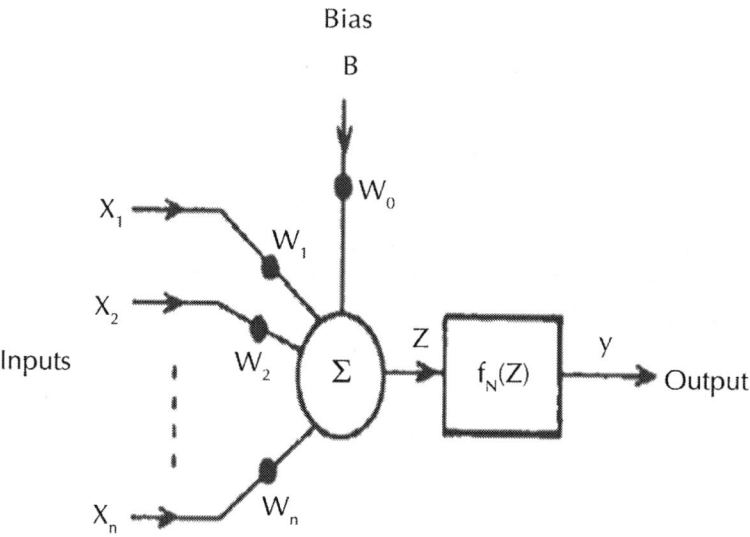

Figure 3: A biased neuron.

The objective of crude oil distillation unit is to perform a process optimization including high production rate with required product quality and low operating costs by searching an optimal operating condition of the operating variables [7]. Genetic algorithm and fuzzy logic are adaptive methods which can be used to solve optimization problems. Genetic Algorithms are based on the principle of genetic processes of biological organisms while Fuzzy logic utilizes human knowledge by giving the fuzzy or linguistic descriptions a definite structure. An expert system design of this unit is necessary in order to achieve the earlier stated goals. An expert system is a computer system employing expert knowledge to attain high levels of performance in solving the problems within a specific domain area [8].

LITERATURE REVIEW

For effective performance and operation of the crude oil distillation column, competent design of the CDU that improve fuel properties, increase the yields of the distillate products and meet environmental specifications is very important. Traditional systems of crude oil distillation column design, optimization and control by previous researchers are discussed in this section. Manley [9] designed crude oil distillation column through the use of improved process thermodynamics which was also used by Santana et al. [10] to assess and improve the energy efficiency of a crude-oil tower. They were limited to the fact that the improved process designs were unproven, projected capital cost reductions were uncertain and energy savings at current prices would not justify the investment needed for process development and commercialization. The design of crude oil distillation column through the use of on-line soft-sensor for its control and optimization was discussed by Yu et al. [11] but they did not consider the design for offline application.

The use of neural network architectures to design refinery crude distillation column for the prediction of product quality was proposed by Bawazir et al. [12]. However, the design of crude oil distillation column for optimization using the neural network was left out. Liau et al. [13] picked up the challenges by using neural network for the expert system design of a crude oil distillation unit for process optimization. Torgashov [14] designed self-optimizing control of complex crude distillation column control using non-linear process model-based method. Khairiyah et al. [15] used the method proposed by Bawazir et al. [12] for design of crude oil distillation column for real time optimization. Gadalla et al. [16] used short-cut models for retrofit design of distillation columns and heat exchanger networks but were limited to the fact that short-cut models were not detailed enough and require longer time for convergence. The stage-wise modelling procedure approach was used for the design and optimization of a refinery crude distillation unit in the context of total energy requirement by Okeke et al. [17]. However, they failed to give full practical validation of the design method. Domijan et al. [18] presented a model that used the crude true boiling point curve and other routinely made laboratory measurements on crude distillation unit (CDU) for product properties prediction.

Another system of crude oil distillation column design was proposed by Macías-Hernández et al. [19] using evolving Takagi-Sugeno fuzzy models for the design of soft sensor for predicting crude oil distillation side streams. However, future research need to be directed at practical implementation of this technique in an online estimator in closed loop. Zalizawati [20] designed a continuous distillation column through development of multiple-input multiple-output (MIMO) and multiple-input single-output (MISO) neural network models. Haydary et al. [21] proposed steady-state and dynamic simulation of crude oil distillation using ASPEN PLUS and ASPEN DYNAMICS. Kanthasamy [22] used Hammerstein model and nonlinear autoregressive model with exogenous input (NARX) in designing nonlinear model predictive control (NMPC) of a distillation column. He included the activity and fugacity calculations in the model in order to account for the non-ideality of the system. Smith et al. [23] proposed design of heat integrated distillation system. However, their work was not detailed enough as the real practical analysis of the system was not stated. Tonnang et al. [3] developed a powerful controller that allows perfect control of a complex crude oil distillation column using neural network. Kansha et al. [24] investigated the crude oil distillation design with exergy analysis and the feasibility for application of the self-heat recuperation technology. The simulation was conducted by PRO/II Ver.8.1 (Invensys, Simsci) to calculate the energy required. However, this design required modification such as a self-heat exchange and an additional furnace heater for the first distillation column and no integration for hot charge method.

RESEARCH METHODOLOGIES

Design and Control of CDU Using ANN

Artificial neural network (ANN) models are black box models, consisting of layers of nodes with nonlinear basis functions and weighted connections that link the nodes. The inputs to the model are mapped to the outputs after being trained with a set of training or learning data to optimise the weights and biases of the nodes. Multilayer feed forward ANN (Figure 4) was mathematically proven to be a universal approximator [25]. The back propagation algorithm proposed by Rumelhart et al. [26] is the effective means for training or learning the set of data in order to optimize the weights. It is an approximate steepest descent algorithm in which performance index is mean square error [27]. Figure 5 gives the implementation procedure for the ANN. The error function of the output is given by:

$$\text{error} = \frac{1}{2}\sum_{j=1}^{m}\left[d_j(n) - y_j(n)\right]$$

(1)

Where d_j = Practical data of jth output neuron;

y_j = Computed data of jth ouput neuron;

m= Neuron number;

n = Training step.

$$y_k = F_N(Z_k) = \frac{1}{1 + \exp(-Z_k)}$$

(2)

Where $y_k = F_N(Z_k)$ = sigmoidal transfer function, Z_k = sum of the jth input to the neuron multiplied by their respective weights

$$Z_k = \sum_j w_{kj} x_j$$

(3)

Where w_{kj} = weight of the jth input to the kth neuron of the output layer; x_j= jth input to the neuron.

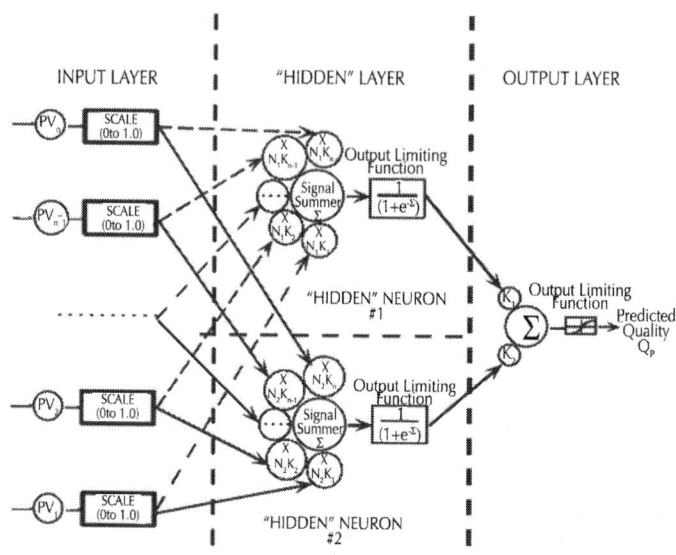

Figure 4: Multilayer feed forward ANN.

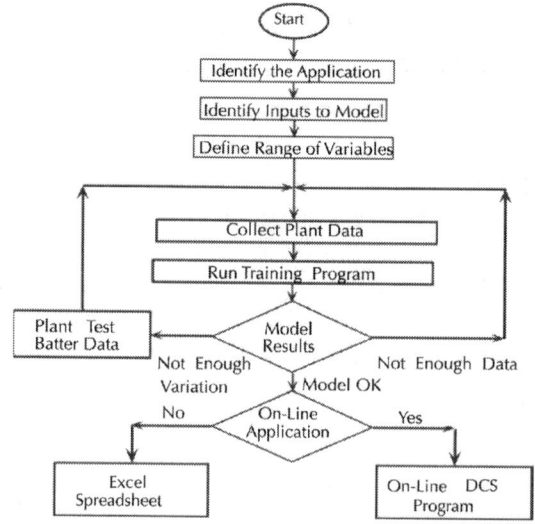

Figure 5: ANN implementation flow chart.

$$w_{ij}^{k+1} = w_{ij}^{k} + \eta \delta_{j}^{k} I_i f'(s)$$

(4)

Where w_{ij}^{k+1} = Weigths of the connection from unit i in layer k + 1 to unit j in layer k + 2;

w_{ij}^{k} = Weigths of the connection from unit i in layer k to unit j in layer k + 1;

η = Learning rate constant;

δ_{j}^{k} = Signal error;

I_i = Input vector to the networks;

$f'(s)$ = Derivative of the networks sigmoidal transfer function;

s = Sum of all the weigths.

The development of neural network that could be used for the control of the crude oil distillation column is also discussed in this section. The tower or column receives crude oil and steam flow as inputs. Naphthalene, Kerosene, Light Diesel Oil and Heavy Diesel Oil are its outputs. Stripping (distillate flows) is sent to the storage tank, while some quantity of Naphthalene, Kerosene and Light Diesel Oil (reflux flows) are returned into the column. The input values to the neural network controller (NNC) are: distillate flows, feed flow, feed temperature, top temperature, bottom temperature, bottom composition, reflux temperature, and the tower pressure. Its output values are used to adjust the reflux flows and steam flow. The neural network controller flow chart is presented in Figure 6.

Optimization of CDU Using Genetic Algorithm and Fuzzy Logic

Genetic Algorithm is a powerful optimization technique based on the principles of natural evolution and selection. In the specific case of selecting the optimum set of inputs from a larger set, GA can be used to search through a large number of input combinations with interdependent variables in the artificial neural network to be designed for the crude oil distillation column. Figure 7 gives the standard structure of genetic algorithm.

Fuzzy theory is another powerful tool in the exploration of complex problems because of its ability to determine outputs for a given set of inputs without using a conventional, mathematical model. The development of fuzzy theory came from the inability to describe some physical phenomena with the exact mathematical models dictated by more conventional Boolean models. Fuzziness describes event ambiguity. It measures the degree to which an event occurs, not whether it occurs. In its simplest form a fuzzy logic is simply a set of rules describing a set of actions to be taken for a given set of inputs. It is easiest to think of these rules as if then statements of the form if {set of inputs} then {outputs}.

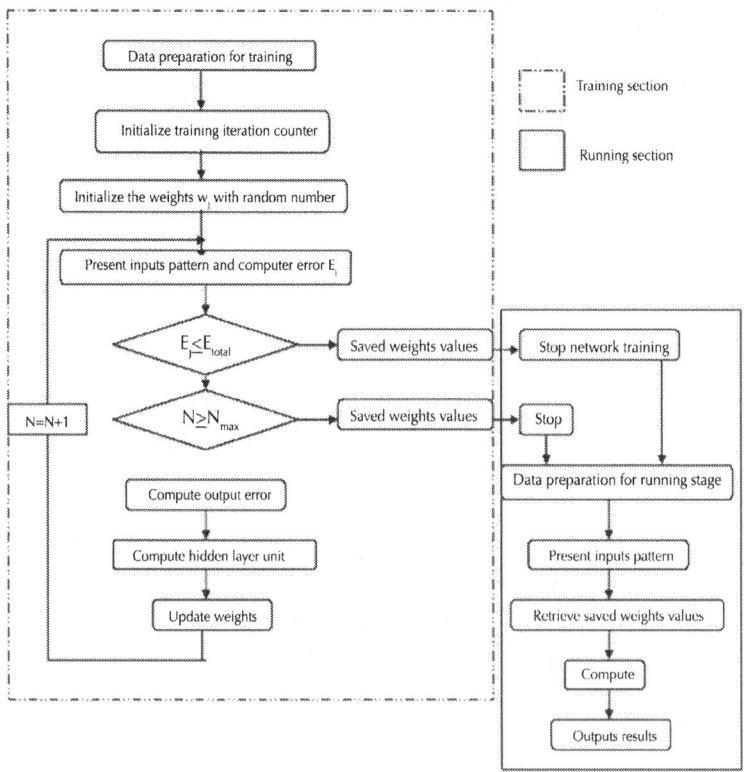

Figure 6: Neural network controller flow chart.

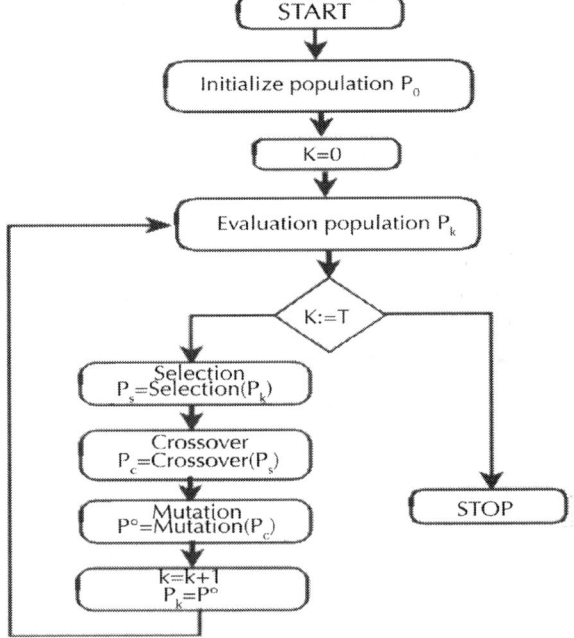

Figure 7: Standard genetic algorithm.

DISCUSSION

After adequate scrutiny of various traditional systems of crude oil distillation column design, optimization and control, artificial neural network, fuzzy logic and genetic algorithm had been found as the best methodologies based on some established facts. They serve as substitutes for dynamic mathematical models as they are time independent. Many researchers used rigorous simulations which led to convergence problems and were also time consuming. These soft computing methodologies use error minimal method to replace the convergence problem. Also, artificial neural network models, fuzzy logic and genetic algorithm approach had been found as the effective ways to model complex processes due to their non-linear characteristic structures. Lastly, the proposed methodologies can remarkably enhance the regulatory and advanced control capabilities of various industrial processes such as crude oil distillation columns in refineries.

CONCLUSIONS

An expert system design of crude oil distillation column can be done using the artificial neural networks. The product quality specification and the optimal operation can be reached through the use of artificial neural network. Also, the crude oil distillation column optimization can be achieved using both the fuzzy logic and genetic algorithm frame work. The continuous evaluation and adjustment of process operating conditions to optimise economic productivity can be reached by these methodologies. A neural network controller can be designed for crude oil distillation column. The developed neural network controller is capable of mapping the interactions and nonlinear dynamics of the process. Artificial neural networks, fuzzy logic and genetic algorithm framework are the best soft computing methodologies for both the expert design and optimization of crude oil distillation column. The design of neural network controller for the crude oil distillation column is also recommended in order to meet the requirements with respect to environment, health and safety of the plant personnel and the quality of the finished products.

REFERENCES

1.   K. H. Bawazeer, "Prediction of Crude Oil Product Quality Parameters Using Neural Networks," MS Thesis, Florida Atlantic University, Boca Raton, 1996.

2.   W. Heather, "Refining Crude Oil," The New Zealand Refining Company Ltd., Ruakaka, 2003.

3.   Z. E. H. Tonnang, "Distillation Column Control Using Artificial Neural Networks," M.Sc Thesis, Microprocessors and Control Engineering, Department of Electrical and Electronics Engineering, Faculty of Technology, University of Ibadan, Ibadan, 2010.

4.   S. Christos and S. Dimitrios, "Introduction to Artificial Neural Network," Research Report, 2001.

5.   H. L. Hoffman, D. E. Lupfer, L. A. Kane and B. A. Jensen, "Distillation Column, Basic and Advance Controls Process Control, Instrument Engineer Handbook," 3rd Edition, Butherworth Heinemann, Oxford, 1995.

6.   T. M. Gowrie and V. V. C. Reddy, "Load Forecasting by a Novel Technique Using ANN," ARPN Journal of Engineering and Applied Sciences, Vol. 3, No. 1, 2008, pp. 19-25.

7.   J. W. Sea, M. Oh and T. H. Lee, "Design Optimization of Crude Oil Distillation," Chemical Engineering Technology, Vol. 23, No. 2, 2000, pp. 157-164. doi:10.1002/(SICI)1521-4125(200002)23:2<157::AID-CEAT157>3.0.CO;2-C

8.   J. McCarthy, "Some Expert System Need Common Sense," Stanford University, Stanford, 1984.

9.   D. B. Manley, "Waste Minimization through Improved Process Thermodynamics: Crude Oil Fractionation," The University of Missouri, Rolla, 1993.

10. E. I. Santana and R. J. Zemp, "Thermodynamic Analysis of a Crude-Oil Fractionating Process," 4th Mercosur Congress on Process Systems Engineering, Vol. 21S, 2001, pp. 523-528.

11. J. J. Yu, C. H. Zhou, S. Tan and C. C. Hang, "An On-line Soft-Sensor for Control and Optimization of Crude Distillation Column," Research Institute of Industrial Process Control, Zhejiang University, Hangzhou, 1997.

12. K. H. Bawazir and A. Zilouchian, "Application of Neural Networks in Oil Refineries," Proceedings of 1996 IEEE International Conference on Neural Networks, New Orleans, 11-13 May 1999, pp. 126-135.

13. L. C.-K. Liau, T. C.-K. Yangb and M. T. Tsaib, "Expert System of a Crude Oil Distillation Unit for Process Optimization Using Neural Networks," Expert Systems with Applications, Vol. 26, No. 2, 2004, pp. 247-255. doi:10.1016/S0957-4174(03)00139-8

14. A. Torgashov, "Nonlinear Process Model-Based Self-Optimizing Control of Complex Crude Distillation Column," European Symposium on Computer Aided Process Engineering-11, Vol. 9, 2001, pp. 793-798.

15. M. F. Khairiyah, K. Fakhri and L. D. Peter, "Connectionist Models of a Crude Oil Distillation Column for Real Time Optimisation," Regional Symposium on Chemical Engineering 2002, Songkla, 2002.

16. M. Gadalla, M. Jobson and M. Smith, "Optimisation of Existing Heat—Integrated Refinery Distillation Systems," Ph.D. Thesis, UMIST, Manchester, 2002.

17. E. O. Okeke and A. A. Osakwe-Akofe, "Optimization of a Refinery Crude Distillation Unit in the Context of Total Energy Requirement," APACT03, 28-30 April 2003, NNPC R&D Division, Port Harcourt, 2003.

18. P. Domijan and D. Kalpić, "Off-Line Energy Optimization Model for Crude Distillation Unit," Ph.D. Thesis, Department of Applied Mathematics, Faculty of Electrical Engineering and Computing, University of Zagreb, Zagreb, 2004.

19. J. J. Macías-Hernández, P. Angelov and X. Zhou, "Soft Sensor for Predicting Crude Oil Distillation Side Streams Using Evolving Takagi-Sugeno Fuzzy Models. Results Outlined," Proceedings of 2nd International Symposium on Evolving Fuzzy Systems, 7-9 September 2008, Lake District, IEEE Press, 2008, pp. 214-220.

20. B. A. Zalizawati, "Development of Multiple-Input MultipleOutput and Multiple-Input Single-Output Neural Network Models for Continuous Distillation Column," M.Sc Thesis, School of Chemical Engineering, Malaysia University, 2008.

21. R. Kanthasamy, "Nonlinear Model Predictive Control of a Distillation Column Using Hammerstein Model and Nonlinear Autoregressive Model with Exogenous Input," Ph.D. Thesis, Universiti Sains Malaysia, 2009.

22. J. Haydary and T. Pavlik, "Steady-State and Dynamic Simulation of Crude Oil Distillation Using ASPEN Plus and ASPEN Dynamics," Petroleum and Coal, Vol. 51, No. 2, 2009, p. 100.

23. R. Smith, M. Jobson, L. Chen and S. Farrokhpanah, "Heat Integrated Distillation System Design," Centre for Process Integration, School of Chemical Engineering and Analytical Science, The University of Manchester, 2010.

24. Y. Kansha, A. Kishimoto and A. Tsutsumi, "Application of the Self-Heat Recuperation Technology to Crude Oil Distillation," Collaborative Research Centre for Energy Engineering, Institute of Industrial Science, The University of Tokyo, Tokyo, 2011.

25. K. Hornik, M. Stinchcombe and H. White, "Multilayer Feedforward Neural Networks Are Universal Approximators," Neural Networks, Vol. 2, No. 5, 1989, pp. 359- 366.doi:10.1016/0893-6080(89)90020-8

26. D. Rumelhart and J. McClelland, "Parallel Distributed Processing," MIT Press, Cambridge, 1986.

27. G. Daniel, "Principles of Artificial Neural Networks," Advanced Series in Circuits and Systems, Vol. 51, No. 2, 2007, pp. 100-109.

Control of Microdomain Orientation in Block Copolymer Thin Films by Electric Field for Proton Exchange Membrane

Joonwon Bae

Department of Applied Chemistry, Dongduk Women's University, Seoul, South Korea

ABSTRACT

Owing to the recent push toward efficient energy storage/conversion devices, fuel cells have become a strong candidate for energy conversion equipments. On the other hand, block copolymer polyelectrolytes are interesting materials for proton exchange membranes in fuel cells. Thus a considerable attention has been paid to the development of block copolymer polyelectrolyte membranes. In this study, the microdomains in block copolymer polyelectrolytes were controlled by external electric fields to develop high performance membranes with improved proton conductivity. The microdomain alignments in sulfonated polystyrene-b-hydrogenated poly butadiene-b-polystyrene block copolymer electrolyte were monitored by cross-sectional transmission electron microscopy analysis. The proton conductivities of the block copolymer electrolyte membranes were measured before and after exposure to electric field. In addition, the morphological features of the block copolymer electrolyte were observed with small angle x-ray scattering and atomic force microscopy.

INTRODUCTION

Recently, energy storage/conversion devices are attracting a considerable amount of interest from all fields of science and technology. These equipments include batteries, capacitors, fuel cells, and solar cells. Among these novel devices, fuel cells are the most strong candidate for the energy conversion systems, which can convert chemical energy directly into electric energy [1] .

In order to improve the overall performances and efficiencies of fuel cells, it is critical to incorporate high performance proton conducting membranes (PCMs), especially in two kinds of popular fuel cells such as proton exchange membrane fuel cell (PEMFC) and direct methanol fuel cell (DMFC) [2] [3] . To date, the PCM based on sulfonated fluorocarbon polymer, for example, Nafion⁻ is the most widely used one [4] . Even if Nafion has high proton conductivity and good thermal stability, its methanol permeability is too high to use in practical applications [4] . Therefore, several attempts have been made in order to develop Nafion⁻ membranes exhibiting reduced methanol permeability while maintaining proton conductivity [4] -[8] .

On the other side, numerous studies have been conducted on developing new PCMs using block copolymers (BCPs) over the past decades [9] -[19] . Attention to the block copolymer polyelectrolytes (BCPEs) as a candidate for PCMs has been paid, because they exhibit unique morphologies such as sphere, cylinder, lamellar, and gyroid due to the self assembled-nanostructures. It has been found that the microdomain structure in BCPEs is strongly dependent on the solvent [20] , annealing condition [21] , shear force [22] , and film thickness [23] . Also, it is very important to understand transport properties such as proton conductivity and methanol permeability through

BCPE membranes when the size, shape, and ordering directions of microdomains in the membranes are different [19] [24] .

Among various BCPEs, the most extensively studied material for PCMs is BCPEs based on the thermoplastic elastomers such as styrene-b-ethylene/butylene-b-styrene (SEBS), which is derived from the hydrogenation reaction of triblock copolymer, styrene-b-butadiene-b-styrene [25] -[39] . The synthesis [25] , properties [27] [28] [31] [32] , and morphologies [37] -[39] of the SEBS BCPEsare well documented. Kim's group has reported the transport properties and morphological characteristics of sulfonated SEBS extensively [10] [38] [39] .

Even if the chemical and physical properties of sulfonated SEBS BCPEs have been extensively investigated, however, the effect of microdomain alignments in SEBS BCPEs on the proton conductivity of PCMs has been scarcely investigated yet [40] . In this study, the microdomain orientations in sulfonated SEBS BCPEs were controlled by external electric fields. The dependence of proton conductivity on microdomain alignments was examined by transmission electron microscopy (TEM) and conductivity measurements. It is certain that this work provides fundamental information for understanding the structure-property relation in BCPE PCMs.

EXPERIMENTAL

Materials

A block copolymer, styrene-b-ethylene/butylene-b-styrene (SEBS, 80 Kg/mol, 28 wt.% polystyrene) was supplied by Kuraray (SEPTON). The block copolymer was sulfonated to prepare sulfonated SEBS (sSEBS) according to the well-established procedure in the previous literature [25] . The sulfonating reagent was acetyl sulfate. The degree of sulfonation was found 22 mol%/hard segment by elemental analysis. Figure 1 exhibits the chemical structure of sSEBS BCPE. A solvent, tetrahydrofuran (THF) was dried by a common method.

Preparation of Thin sSEBS Films

A solution of sSEBS in THF was prepared and filtered. Concentration of the solution was 3 wt.%. A sSEBS film was spin-coated onto silicon wafer at 2000 rpm for 60 s from the solution. Then the films were dried under ambient conditions until complete evaporation of solvent.

Preparation of Thick sSEBS Films for SAXS

A 10 wt.% solution of sSEBS in THF was prepared and poured into a glass dish. Then the film was placed at room temperature until complete evaporation of solvent. The membrane was recovered as a flat-sheet type and the thickness was measured at dry state using a digital micrometer.

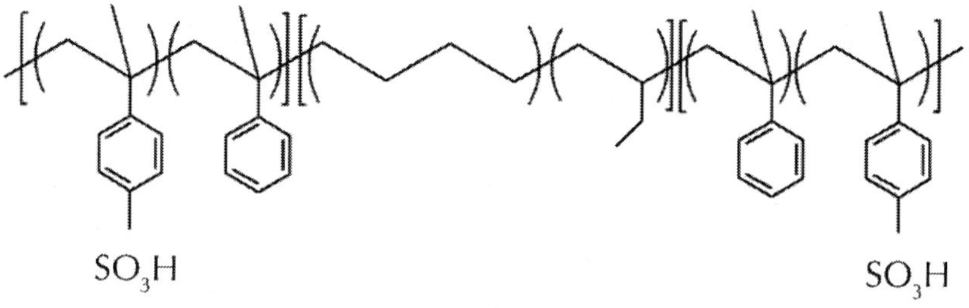

SO$_3$H   SO$_3$H

Figure 1: Chemical structure of the sulfonated SEBS electrolyte.

Application of Electric Field

Figure 2 shows the diagram of experimental apparatus for application of electric field to the prepared sSEBS film. An aluminized Kapton film served as the top electrode and asilicon wafer covered with Al foil as the other. A thin layer (20 - 25 mm) of cross-linked PDMS (Sylgard) was used as a buffer layer between the Kapton electrode and the copolymer thin film. The PDMS layer conforms to the electrode surface, eliminates air gaps between the top electrode and the copolymer film, and maintains asmooth surface of the copolymer film. The copolymer films were heated to 130°C under N$_2$ with an applied electric fieldof 20 V/mm for 6 ~ 12 h and then quenched to room temperature before removing the electric field.

Proton Conductivity and Water Uptake Measurement

The membranes were immersed in deionized water at least 6 h before measurements.Proton conductivity was measured using a complex impedance analyzer (ZAH-NER IM-6). AsSEBS membrane was sandwiched between two stainless electrodes and an ac perturbation of 1 V was applied to the cell.

The membranes were soaked for at least 3 days in deionized water and weighed to determine the uptake contents. Weight of a dried membrane was measured after drying the sample at 60°C under vacuum overnight. The uptake content was calculated by

$$\text{Uptake content}\,(\%) = \frac{w_{wet} - w_{dry}}{w_{dry}} \times 100$$

(1)

where w_{dry} and w_{wet} are the mass of dried and wet samples, respectively.

Instrumentations

Small angle x-ray scattering (SAXS) patterns were obtained with a Bruker AXS Nanostar small angle X-ray scattering spectrometer with a generator voltage of 40 KV and a current of 35 mA. Scanning force microscopy (SFM) images were obtained in both the height and phase-contrast mode using a Digital Instruments Dimension 3100 scanning force microscope in a tapping mode. Transmission electron microscopy (TEM) experiments were performed with a JEOL 200CX TEM operated at an accelerating voltage of 200 kV. To prepare cross-sectional TEM specimens, a thin layer of carbon was evaporated onto the film surface before embedding in epoxy resin to prevent the diffusion of the epoxy resin into hybrid thin film. The thin film was then embedded in an epoxy resin and cured at 60°C for 12 h. The films were removed from the substrate by dipping into liquid N_2. Ultrathin sections (60 nm) were collected at room temperature using a Leica Ultracut Microtome, equipped with a diamond knife. The thin sections were exposed to RuO_4 for 10 min to enhance the contrast.

RESULTS & DISCUSSION

In determining sulfonation degree from the elemental analysis result, it must be assumed that the sulfonation takes place exclusively on the polystyrene end blocks. It should be mentioned that the sulfonation degree actually obtained is significantly lower than the expected value [25] . The efficiency of sulfonation employed in this study is approximately 22%.

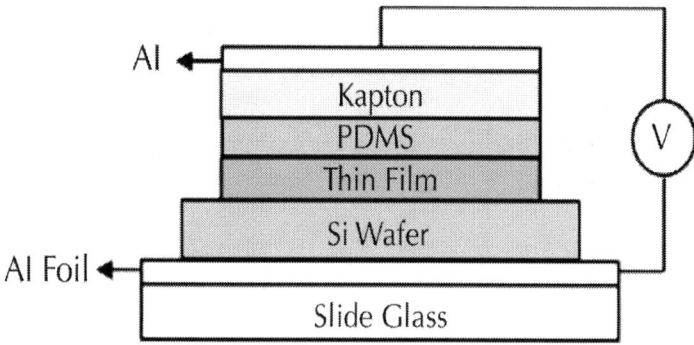

Figure 2: Schematic diagram of the experimental apparatus for application of electric field.

In addition, it is well known that proton conducts through the ionic channels or the water clusters where water or methanol molecules are likely to swell a polymer matrix. The membranes prepared in this study have hydrophilic sulfonic acid groups enough to show the high proton conductivity because of their morphologies with percolated ionic channels [10] . In general, the proton conductivity is primarily dependent on the concentration of sulfonic acid groups. More specifically, the proton conductivity is closely related to the average number of water molecules per fixed sulfonic acid group [39] . Therefore, the water uptake content was measured for sSEBS membranes using Equation (1). The obtained water uptake for pristine sSEBS film was approximately 20 wt.% regardless of film thickness. It was also revealed that the exposure to electric field did not significantly change the water uptake amount.

Figure 3 illustrates SAXS profiles of sSEBS film cast from THF before (solid line) and after (dotted line) thermal annealing at 130°C under N_2 for 48 h. THF was selected as solvent in this experiment because it is a good solvent for both constituent blocks. In addition, the molecular architecture of sSEBS in THF was clearly examined [28] . The thermal annealing has been performed well above the glass transition temperature (T_g) of both blocks, 100 and −40°C for PS and PEB, respectively. It is shown that the sSEBS membrane have lamellar microdomain morphology as indicated by the presence of maxima at positions with ratios of 1:1.95. The peak positions up to the second correlation peak are well resolved for the curve. It is known that the microscopically arranged lamellae of block copolymers exhibit the position of maximum intensity at q values o 1, 2, 3, 4, etc. It was also found that the apparent spacing (D) by using Bragg's equation, D = $2p/q_{max}$, was estimated to be 35.1 nm from the position of first-order peak. It is accepted that the d-spacing is remarkably influenced by the solvent, water uptake, and sulfonation degree [38] [39] .

Figure 4 displays the tapping mode phase image of the sSEBS sample. The unique morphology consisting of bright and dark microdomain is observed remarkably. For sSEBS (14/72/14), an intermediate between the cylinder and lamellar structure is formed. It has been reported that a further slight increase in PS weight ratio changed the microstructure to a well-ordered lamellar [37] . The obtained morphology appears worm-like with lamellar regions dispersed throughout, and might be termed "frustrated" due to the relative disorder [41] .

Figure 5 exhibits the cross-sectional TEM images of sSEBS membrane obtained before and after exposure to electric field (20 V/mm). Figure 5(a) shows the equilibrium microstructure of sSEBS after thermal annealing at 130°C under N_2 for 48 h. The microdomain morphology is apparently lamellar throughout the entire film, where the bright and dark regions correspond to elastomer and polystyrene, respectively. This phenomenon can be explained by the following reasoning.

As for block copolymer thin film, the mean field theory (including Flory-Huggins) offers a fundamental concept of phase separation. In particular, the lamellar phases of triblock copolymer are well-described by strong segregation theory [42] . It has been shown that the domain spacing (D) for the lamellar structures of block copolymer is dependent on three physical parameters such as volume fraction of chain (f), molecular weight (N), and interaction parameters (c) between blocks when the morphology shows lamellar phases. However, it is often difficult to evaluate the exact values of interaction parameters, because the three parameters are strong dependent on temperature and compositions. Instead of interaction parameters, the solubility parameters are of use to understand the interaction between polymers. In particular, the solubility parameter is a useful measure of interaction between blocks, when polymers are partially functionalized with a group as seen in the BCPE employed in this study. Alternatively, the interaction parameter is given by

$$\chi_{sPS-EB} = \left(\delta_{sPS} - \delta_{EB}\right)^2 \big/ \rho_0 k_b T \tag{2}$$

where d_i is the solubility parameter of block i, r_0 the number density of the mono-unit, and k_b the Boltzmann constant [43] . The reference values for d_{PS}, d_{sPS}, and d_{EB} are 9.1, 16.6, and 7.8 $cal^{1/2}/cm^{2/3}$, respectively [44] . It was calculated that the value of cN for the sSEBS sample is approximately 200 at room temperature. According to this value, the morphology of sSEBS is likely to form a lamellar structure, as confirmed by Figure 5(a). It is commonly accepted that the cN values were influenced by the intermolecular ionic aggregation within the partially sulfonated polystyrene domains. Furthermore, strong ionic interactions like hydrogen bondings play role of physical crosslinking, which might induce an effect of increasing molecular weight.

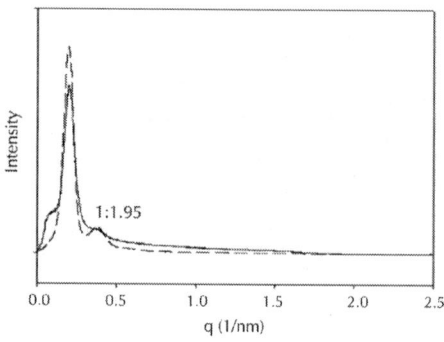

Figure 3: Small angle x-ray scattering patterns of the sulfonated SEBS membrane having an equilibrium microstructure obtained before (solid) and after (dotted) thermal annealing at 130°C under N$_2$ for 48 h.

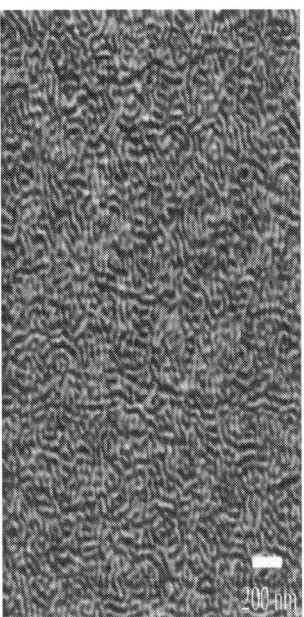

Figure 4: Scanning force microscopy phase image of the as-spun sulfonated SEBS membrane obtained at tapping mode.

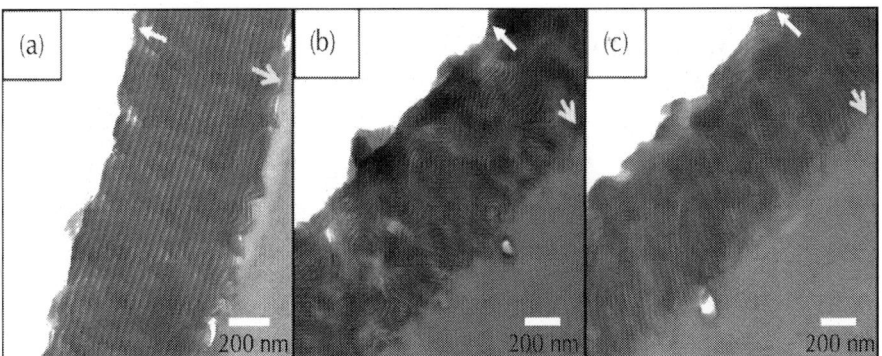

Figure 5: Cross-sectional TEM images of the sulfonated SEBS membranes (a) thermally annealed at 130°C under N_2 for 48 h, exposed to electric field of 20 V/mm for (b) 6 h and (c) 12 h at 130°C under N_2. The white and yellow arrows indicate the film/substrate and the film/air interfaces, respectively.

Figure 5(b) and Figure 5(c) present the cross-sectional TEM images of sSEBS films exposed to an electric field of 20 V/mm as a function of exposure time. It is conspicous that the microdomains oriented normal to the surface, adapted to the direction of applied electric field (yellow arrow). When an applied electric field exceeds a critical value to overcome interfacial interaction, the surface/interface becomes unstable. The electric field aligns interfaces separating two dielectric blocks parallel to the electric field vector. This leads to the formation of arrays of microdomains normal to the surface to minimize the moment of electrostatic forces exerted onto the interface between two blocks, driven by differences in the dielectric constants of the two bodies at the interface [45]

It is also discernible that the alignments started locally after 6 h exposure and then extended to the overall range in the middle of BCP thin film under an electric field. It is notable that the vertical alignment extends completely to the film/substrate interface (white arrow). This is a strikingly different phenomenon from that seen in most block copolymer thin films. This is because the presence of ionic functional groups in polystyrene block can promote the response of PS microdomain to the direction of electric field even at the vicinity of film/substrate interface. In this case, the electrostatic force on sulfonated polystyrene domain exerted by the electric field might easily surpass the interfacial energy [46]

Table 1: Summary of the proton conductivities of the membranes

Sample	Proton conductivity (S/cm)
As spun (<1 µm)	0.018
Thin film (<1 µm) exposed to e-field	0.021
Thick film (30 µm) exposed to e-field	0.029

The proton conductivities of sSEBS samples are summarized in Table 1. The proton conductivity is intimately dependent on the various parameters such as sulfonation degree, water uptake, and methanol permeability. In this experiment, the effect of those parameters is excluded, because the other experimental parameters are fixed. For thin sSEBS membranes (<1 mm), the proton conductivity slightly increased after application of electric field. On the contrary, the conductivity increased significantly by exposure to electric field for thick membrane. This is attributed to the facilitated formation of percolated conducting channels inside membrane through microdomain alignments. Meanwhile, the increased water uptake amount in thicker membrane might be of help to elevate the proton conductivity. However, the mere application of electric field did not elevate the proton conductivity significantly. This is because conductivity requires connectivity of microdomainsacross macroscopic length scales and obtaining this in the aligned orientation appears difficult, especially for thin films.

CONCLUSIONS

The dependence of proton conductivities on microdomain alignments in sulfonated SEBS block copolymer electrolytes was revealed. It could be concluded that the proton conductivities of sulfonated SEBS membranes were elevated by the application of electric field to promote microdomain alignments. In addition, a schematic pathway to microdomain orientation in sulfonated SEBS under electric field was presented. A more detailed study regarding this topic is underway. This article can provide an interesting outcome for understanding the structure-property relation in proton exchange membranes.

ACKNOWLEDGEMENTS

This study was supported by Dongduk Women's University Grant [2012-02204].

REFERENCES

1.   Baldauf, M. and Preidel, W. (1999) Status of the Development of Direct Methanol Fuel Cell. Journal of Power Sources, 84, 161-166. http://dx.doi.org/10.1016/S0378-7753(99)00332-8.

2.   Rikukawa, M. and Sanui, K. (2000) Proton-Conduction Polymer Electrolyte Membranes Based on Hydrocarbon Polymers. Progress in Polymer Science, 25, 1463-1502.http://dx.doi.org/10.1016/S0079-6700(00)00032-0.

3.   Elabd, Y.A. and Hickner, M.A. (2011) Block Copolymers for Fuel Cells. Macromolecules, 44, 1-11. http://dx.doi.org/10.1021/ma101247c.

4.   Heitner-Wirguin, C. (1996) Recent advances in Perfluorinated Ionomer Membranes: Structure, Properties and Applications. Journal of Membrane Science, 120, 1-33.http://dx.doi.org/10.1016/0376-7388(96)00155-X.

5.   Costamagna, P. and Srinivasan, S. (2001) Quantum Jumps in the PEMFC Science and Technology from the 1960s to the Year 2000; Part I. Fundamental Scientific Aspects. Journal of Membrane Science, 102, 242-252.

6.   Vielstich, W. and Ives, D.J.G. (1970)Fuel cells; Modern processes for the Electrochemical Production of Energy. Wiley, London.

7.   Tricoli, V. (1998) Proton and Methanol Transport in Poly (Perfluorosulfonate) Membranes Containing Cs^+ and H^+ Cations. Journal of the Electrochemical Society, 145, 3798-3801.http://dx.doi.org/10.1149/1.1838876

8.   Tricoli, V., Carretta, N. and Bartolozzi, M. (2000) A Comparative Investigation of Proton and Methanol Transport in Fluorinated Ionomeric Membranes. Journal of the Electrochemical Society, 147, 1286-1290. http://dx.doi.org/10.1149/1.1393351

9.   Hickner, M.A., Ghassemi, H., Kim, Y.S., Einsla, B.R. and McGrath, J.E. (2004) Alternative Polymer Systems for Proton Exchange Membranes (PEMs). Chemical Reviews, 104, 4587-4612. http://dx.doi.org/10.1021/cr020711a

10.   Kim, J., Kim, B. and Jung, B. (2002) Proton Conductivities and Methanol Permeabilities of Membranes Made from Partially Sulfonated Polystyrene-Block-Poly (Ethylene-Ran-Butylene)-Block-Polystyrene Copolymers. Journal of Membrane Science, 207, 129-137.http://dx.doi.org/10.1016/S0376-7388(02)00138-2

11.   Elabd, Y.A., Napadensky, E., Sloan, J.M., Crawford, D.M. and Walker, C.W. (2003) Triblock Copolymer Ionomer Membranes:

Part I. Methanol and Proton Transport. Journal of Membrane Science, 217, 227-242. http://dx.doi.org/10.1016/S0376-7388(03)00127-3

12. Shi, Z.Q. and Holdcroft, S. (2005) Synthesis and Proton Conductivity of Partially Sulfonated Poly ([Vinylidene Difluoride-co-Hexafluoropropylene]-b-Styrene) Block Copolymers. Macromolecules, 38, 4193-4201. http://dx.doi.org/10.1021/ma0477549

13. Park, M.J., Downing, K.H., Jackson, A., Gomez, E.D., Minor, A.M., Cookson, D., Weber, A.Z. and Balsara, N.P. (2007) Increased Water Retention in Polymer Electrolyte Membranes at Elevated Temperatures Assisted by Capillary Condensation. Nano Letters, 7, 3547-3552. http://dx.doi.org/10.1021/nl072617l

14. Xu, K., Li, K., Khanchaitit, P. and Wang, Q. (2007) Synthesis and Characterization of Self-Assembled SulfonatedPoly (Styrene-b-Vinylidene Fluoride-b-Styrene) Triblock Copolymers for Proton Conductive Membranes. Chemistry of Materials, 19, 5937-5945. http://dx.doi.org/10.1021/cm071626s

15. Saito, T., Moore, H.D. and Hickner, M.A. (2010) Synthesis of Midblock-Sulfonated Triblock Copolymers. Macromolecules, 43, 599-601.http://dx.doi.org/10.1021/ma9023125

16. Chen, L., Hallinan, Jr., D.T., Elabd, Y.A. and Hillmyer, M.A. (2009) Highly Selective Polymer Electrolyte Membranes from Reactive Block Polymers. Macromolecules, 42, 6075-6085. http://dx.doi.org/10.1021/ma901272s

17. Lee, H.-S., Roy, A., Lane, O., Dunn, S. and McGrath, J.E. (2008) Hydrophilic-Hydrophobic Multiblock Copolymers Based on Poly (Arylene Ether Sulfone) via Low-Temperature Coupling Reactions for Proton Exchange Membrane Fuel Cells. Polymer, 49, 715-723. http://dx.doi.org/10.1016/j.polymer.2007.12.023

18. Bae, B., Miyatake, K. and Watanabe, M. (2010) Sulfonated Poly (Arylene Ether Sulfone Ketone) Multiblock Copolymers with Highly Sulfonated Block. Synthesis and Properties. Macromolecules, 43, 2684-2691. http://dx.doi.org/10.1021/ma100291z

19. Csernica, J., Baddour, R.F. and Cohen, R.E. (1987) Gas Permeability of a Polystyrene-Polybutadiene Block Copolymer with Oriented Lamellar Domains. Macromolecules, 20, 2468-2471. http://dx.doi.org/10.1021/ma00176a024

20. Brinkmann, S., Stadler, R. and Thomas, E.L. (1998) New Structural Motif Inhexagonally Ordered Cylindrical Ternary (ABC) Block Copolymer Microdomains. Macromolecules, 31, 6566-6574. http://dx.doi.org/10.1021/ma980103q

21. Hajduk, D.A., Gruner, S.M., Rangarajan, P., Register, R.A., Fetters, L.J., Honeker, C., Albalak, R.J. and Thomas, E.L. (1994) Observation of a Reversiblethermotropic Order-Order Transition in a Diblock Copolymer. Macromolecules, 27, 490-501.http://dx.doi.org/10.1021/ma00080a024

22. Morrison, F.A. and Winter, H.H. (1989) The Effect of Unidirectionalshear on the Structure of Triblock Copolymers. I. Polystyrene-Polybutadiene-Polystyrene. Macromolecules, 22, 3533-3540. http://dx.doi.org/10.1021/ma00199a006

23. Heck, B., Arends, P., Ganter, M., Kressler, J. and Stühn, B. (1997) SAXS and TEM Studies on Poly(strene)-blockpoly(ethene-co-but-1-ene)-block-poly(styrene) in Bulk and at Various Interfaces. Macromolecules, 30, 4559-4566.http://dx.doi.org/10.1021/ma9617072

24. Csernica, J., Baddour, R.F. and Cohen, R.E. (1989) Gas Permeability of Apolystyrene/Polybutadiene Block Copolymer Possessing a Misoriented Lamellar Morphology. Macromolecules, 22, 1493-1496. http://dx.doi.org/10.1021/ma00193a085

25. Weiss, R.A., Sen, A., Willis, C.L. and Pottick, A. (1991) Block Copolymer Ionomers: 1. Synthesis and Physical Properties of Sulfonated Poly(styrene-ethylene/butylene-styrene). Polymer, 32, 1867-1874.

26. Jacobs, P.M. and Jones, F.R. (1989) Diffusion of Moisture into Two-Phase Polymers. Journal of Materials Science, 24, 2331-2336. http://dx.doi.org/10.1007/BF01174492

27. Weiss, R.A., Sen, A., Pottick, A. and Willis, C.L. (1991) Block Copolymer Ionomers: 2.Viscoelastic and Mechanical Properties of Sulfonated Poly(styrene-ethylene/butylene-styrene). Polymer, 32, 2785-2792. http://dx.doi.org/10.1016/0032-3861(91)90109-V

28. Wu, C., Woo, K. and Jiang, M. (1996) Light-Scattering Studies of Styrene-(ethylene-co-butylene)-Styrene Triblock Copolymer and Its Sulfonated Ionomers in Tetrahydrofuran. Macromolecules, 29, 5361-5367. http://dx.doi.org/10.1021/ma960247+

29. Adams, J.L., Quiram, D.J., Graessley, W.W., Register, R.A. and Marchand, G.R. (1998) Interaction Strengths in Styrene-Diene Block Copolymers and Their Hydrogenated Derivatives. Macromolecules, 31, 201-204. http://dx.doi.org/10.1021/ma9710500

30. Ovejero, G., Perez, P., Romero, M.D., Guzman, I. and Diez, E. (2007) Solubility and Flory Huggins Parameters of SBES, Poly(Styrene-b-butene/ethylene-b-styrene) Triblock Copolymer, Determined by Intrinsic Viscosity. European Polymer Journal, 43, 1444-1449.http://dx.doi.org/10.1016/j.eurpolymj.2007.01.007

31. Blackwell, R.I. and Mauritz, K.A. (2004) Dynamic Mechanical Properties of Annealed Sulfonated Poly(styrene-b- [ethylene/butylene]-b-styrene) Block Copolymers. Polymer, 45, 3457-3463. http://dx.doi.org/10.1016/j.polymer.2004.02.010

32. Chen, H., Hassan, M.K., Peddini, S.K. and Mauritz, K.A. (2011) Macromolecular Dynamics of Sulfonated Poly(Styrene-b-ethylene-ran-butylene-b-styrene) Block Copolymers by Broadband Dielectric Spectroscopy. European Polymer Journal, 47, 1936-1948. http://dx.doi.org/10.1016/j.eurpolymj.2011.07.005

33. Han, C.D., Chun, S.B., Hahn, S.F., Harper, S.Q., Savickas, P.J., Meunier, D.M., Li, L. and Yalcin, T. (1998) Phase Behavior of Polystyrene/Polybutadiene and Polystyrene/Hydrogenated Polybutadiene Mixtures: Effect of the Microstructure of Polybutadiene. Macromolecules, 31, 394-402. http://dx.doi.org/10.1021/ma971309e

34. Wang, Y., Shen, J.S. and Long, C.F. (2001) The Effect of Casting Temperature on Morphology of Poly(Styrene-ethylene/butylene-styrene) Triblock Copolymer. Polymer, 42, 8443-8446. http://dx.doi.org/10.1016/S0032-3861(01)00325-1

35. Wang, L., Hong, S., Hu, H., Zhao, J. and Han, C.C. (2007) The Surface Morphology Evolution of an Ultrathin Film of Poly[styrene-b-(ethylene-co-butylene)-b-styrene] during Its Dewetting Process. Langmuir, 23, 2304-2307. http://dx.doi.org/10.1021/la063314u

36. Wang, L., Zhao, J. and Han, C.C. (2008) Phase Separation of Polystyrene-b-(Ethylene-co-butylene)-b-styrene (SEBS) Deposited on Polystyrene Thin Films. Polymer, 49, 2153-2159.

37. Wang, D., Fujinami, S., Liu, H., Nakajima, K. and Nishi, T. (2010) Investigation of True Surface Morphology and Nanomechanical

Properties of Poly(styrene-b-ethylene-co-butylene-b-styrene) Using Nanomechanical Mapping: Effects of Composition. Macromolecules, 43, 9049-9055. http://dx.doi.org/10.1021/ma100959v

38.   Kim, J., Kim, B., Jung, B., Kang, Y.S., Ha, H.Y., Oh, I.H. and Ihn, K.J. (2002) Effect of Casting Solvent on Morphology and PhysicalProperties of Partially Sulfonated Polystyrene-block-poly(ethylene-ran-butylene)-block-polystyrene   Copolymers. Macromolecular Rapid Communications, 23, 753-756. http://dx.doi.org/10.1002/1521-3927(20020901)23:13<753::AID-MARC753>3.0.CO;2-G

39.   Kim, B., Kim, J. and Jung, B. (2005) Morphology and Transport Properties of Protons and Methanol through Partially Sulfonated Block Copolymers. Journal of Membrane Science, 250, 175-182. http://dx.doi.org/10.1016/j.memsci.2004.10.025

40.   Park, M.J. and Balsara, N.P. (2010) Anisotropic Proton Conduction in Aligned Block Copolymer Electrolyte Membranes at Equilibrium with Humid Air. Macromolecules, 43, 292-298. http://dx.doi.org/10.1021/ma901980b

41.   Mauritz, K.A., Blackwell, R.I. and Beyer, F.L. (2004) Viscoelastic Properties and Morphology of Sulfonated Poly(styrene-b-ethylene/butylene-b-styrene) Block Copolymers (sBCP), and sBCP/[silicate] Nanostructured Materials. Polymer, 45, 3001-3016.http://dx.doi.org/10.1016/j.polymer.2003.12.078

42.   Matsen, M.W. (2000) Equilibrium Behavior of Asymmetric ABA Triblock Copolymer Melts, Journal of Chemical Physics, 113, 5539. http://dx.doi.org/10.1063/1.1289889

43.   Shibayama, M., Hashimoto, T. and Kawai, H. (1983) Ordered Structure Inblock Polymer Solutions. 5. Equilibrium and Nonequilibrium Aspects of Microdomain Formation. Macromolecules, 16, 1434-1443. http://dx.doi.org/10.1021/ma00243a006

44.   Krevelen, V. (1990) Properties of Polymers. 3rd Editon, Elsevier, New York, 189.

45.   Xu, T., Hawker, C.J. and Russell, T.P. (2003) Interfacial Energy Effects on the Electric Field Alignment of Symmetric Diblock Copolymers. Macromolecules, 36, 6178-6182.http://dx.doi.org/10.1021/ma034511s

46.   Wang, J.Y., Chen, W., Roy, C., Sievert, J.D. and Russell, T.P. (2008) Influence of Ionic Complexes on Phase Behavior of Polystyrene-b-poly(methyl methacrylate) Copolymers. Macromolecules, 41, 963-969. http://dx.doi.org/10.1021/ma071908d

Chapter 9

Correlating Laminar Burning Velocities using Perfectly Stirred Reactor Theory

Robert B. Barat

Department of Chemical Engineering, Chemistry and Environmental Science, University Heights, New Jersey Institute of Technology, Newark, NJ 07102-1982, USA

ABSTRACT

The laminar burning velocity (LBV) of a gaseous fuel/oxidant feed is a strong function of the feed mixture composition. In this paper, published LBV values are successfully correlated with the square root of the ratio of the thermal diffusivity of the fuel/oxidant feed mixture to the space time at extinction of a hypothetical perfectly stirred reactor (PSR) burning this feed. The PSR simulations use detailed reaction mechanisms. A wide range of hydrocarbon and inorganic fuels and oxidants are considered.

INTRODUCTION

The increasing availability of detailed reaction mechanisms supports a growing interest in the engineering of combustion chemistry, especially regarding pollutants, materials, inhibitors, etc. Combustion is characterized by a complex, non-linear interaction of transport with reaction. Therefore, combustion makes for an excellent application of chemical reaction engineering.

In this paper, published values of laminar burning velocity (LBV) of numerous gaseous fuel/oxidant mixtures are correlated with a calculated quantity based on the thermal diffusivity of the feed mixture and the space time at extinction (blowout) of a hypothetical perfectly stirred reactor (PSR). The blowouts are calculated using published detailed reaction mechanisms. This application of reaction engineering to combustion offers a relatively simple way to estimate LBVs for fuel/oxidant pairs assuming reaction mechanisms are available.

LAMINAR BURNING VELOCITY

A simple thermal treatment of the propagation of a laminar combustion wave through a gaseous fuel/oxidant mixture was considered by Mallard and Le Chatelier in the 19th century. As described by Glassman (1987), the incoming fuel/oxidant mixture is heated in the preheat zone by conduction from the reaction zone. Using a linear approximation to the temperature gradient in the reaction zone, a simple energy balance yields:

$$k\frac{T_b - T_1}{l_r} = mC_p(T_1 - T_u),$$

(1)

where k is the gas thermal conductivity, T_b the burned gas temperature, T_1 an ignition temperature, m the mass flow rate, T_u the unburned gas temperature, C_p a mean gas heat capacity, and l_r the reaction zone thickness, which can be given by: $l_r = v_u t_r$, where t_r is the time in the reaction zone, and v_u the velocity of the unburned gases.

Since the flow cross section can be taken as constant, continuity provides that $m = \rho_u v_u$, where ρ is the gas density. By definition, the laminar burning velocity $S_L = v_u$. Substitution into Eq. (1) yields

$$S_L = \left[\frac{k}{\rho_u C_p} \frac{T_b - T_1}{T_1 - T_u} \frac{1}{t_r} \right]^{0.5}.$$

(2a)

The group $(k/\rho C_p)$ is the thermal diffusivity a. The quantity $1/t_r$ is a characteristic reaction rate R. The exponential temperature dependence of R dominates the linear temperature term. The simple thermal analysis, then, holds that the LBV is proportional to the square root of the product of the thermal diffusivity of the feed gas mixture and the combustion reaction rate:

$$S_L \propto [a\mathcal{R}]^{0.5}.$$

(2b)

The more complex, control volume analysis of Semenov (Glassman, 1987) considers both thermal and species (molecular) diffusion. However, this approach can be simplified to the thermal analysis result of Eq. (2b) (Glassman, 1987) Neither approach considers the important effects of free radicals. Other, more complicated, approaches have been put forth. However, the thermal theory result of Eq. (2b) will be used here as it offers a potentially simple basis for an engineering correlation of a very complex phenomenon.

The LBV of a gaseous fuel/oxidant mixture is a good indicator of its relative combustion rate and stability. Burning velocities can be measured by several techniques (Fristrom, 1995). The simplest is the Bunsen method where the LBV is the product of the cold gas velocity exiting the burner tube and the sine of the angle that the premixed inner core of the flame makes with the vertical axis.

The LBV varies with fuel equivalence ratio[1] (φ), typically reaching a maximum at φ approximately one for hydrocarbon/air systems (Fristrom, 1995). Burning velocities are also quite dependent on levels of dopants, such as halogens. For example, CH_4/air flames doped with C_1 chlorocarbons show a marked decrease in maximum burning velocities as the chlorine content of the fuel increases (Gupta & Valeiras, 1984). Since the LBV reflects the rate of propagation of a laminar combustion wave, it is very useful in correlating the relative strengths of inhibitors and flame retardants (Westbrook, 1983).

In order to utilize Eq. (2b) to correlate published LBV values, a convenient method is needed to estimate the reaction rate R values. Stirred reactor theory will be used for this purpose.

PERFECTLY STIRRED REACTOR

In a PSR, the feed is instantaneously and continuously mixed with the reacting fluid. The utility of the PSR lies with its assumption of infinitely fast mixing relative to chemical rates. This simplification can be challenged under combustion conditions since reaction rates often approach mixing rates at high temperatures. However, practical laboratory devices have been developed (Nenniger, Kridiotis, Chomiak, Longwell, & Sarofim, 1984) which emulate PSR behavior under most conditions of interest in combustion (Barat, 1992). These devices have been successfully employed in many combustion chemistry studies (e.g. Brouwer, Longwell, Sarofim, Barat, & Bozzelli, 1992; Mao & Barat, 1996).

Borrowing the chemical reaction engineering notation applied by Fogler (1999), the respective PSR species and energy balances are

$$F_{Ao}X_A = (-r_A)V, \qquad (3a)$$

$$UA(T_a - T) - F_{Ao}X_A(\Delta H_r) = F_{Ao}\sum_{i=1}^{n} \Theta_i \int_{T_0}^{T} C_{p_i}\, dT, \qquad (3b)$$

Where F_{Ao} is the molar feed rate of fuel A, X_A the conversion of A, $-r_A$ the net molar rate of reaction of A, V the reactor volume, U the overall heat transfer coefficient, T_a the ambient temperature, T the reactor temperature, ΔH_r the heat of reaction, n the number of species, Θ_i the molar flow rate ratio of species i to A in the feed, C_{pi} the specific heat of i, A the heat transfer area, and T_0 the feed temperature.

For illustration purposes, the above problem can be simplified with the following arbitrary assumptions:
- Adiabatic: $UA=0$.
- Reaction: A+B→C+D, elementary kinetics, gas phase.
- Constant heat of reaction, constant heat capacities.
- Stoichiometric feed of A and B: $\Theta_A=1; \Theta_B=1$.
- Constant pressure.

For this problem, Eq. (3) becomes

$$F_{Ao}X_A = A_f \exp(-E/RT)C_{Ao}^2(1 - X_A)^2(T_0/T)^2, \qquad (4a)$$

$$T = T_0 + \frac{X_A(-\Delta H_r)}{C_{pA} + C_{pB}},$$

(4b)

Where A_f, E are the Arrhenius rate constant parameters, and C_{Ao} the molar feed concentration of A.

Eq. (4) was evaluated for an exothermic reaction with the following arbitrary parameters: T_o =300 K, A_f= 10^{12}cm³/mol s, E=29 805 cal/mol, C_{Ao}=2*10^{-5} mol/ cm³, ΔH_r=-50 kcal/mol, F_{Ao}=0.01 mol/s, C_{pA}=20 cal/ mol K, and C_{pB}=10 cal/mol K. The high-temperature solution (X_A=0.947, T=1880 K) represents the stable combustor operating condition for this feed rate, and is shown in Fig. 1.

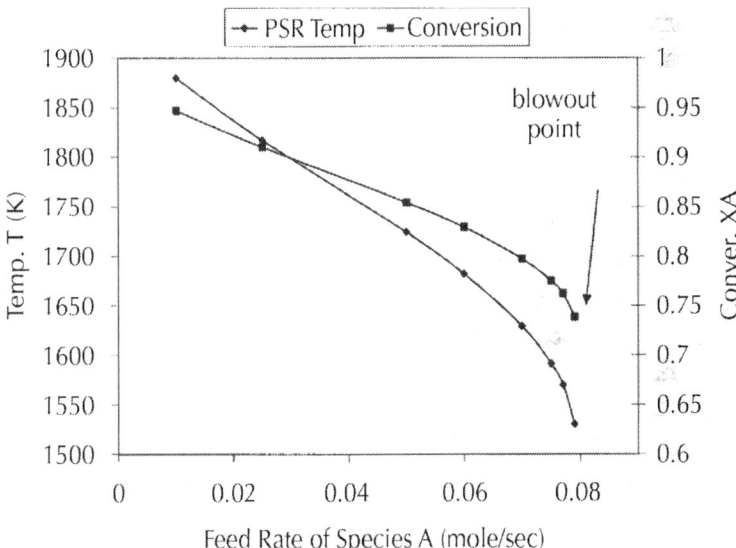

Figure 1: High-temperature stable solutions for simple PSR example with corresponding conversions; extinction (blowout) occurs at approximately

$$F_{Ao} = 0.079 \text{mol}/\text{s}, T = 1530\text{K}, X_A = 0.738.$$

As total feed rate (proportional to F_{Ao}) at constant composition is increased in this problem, the high-temperature solution shifts to lower temperature and conversion values (Fig. 1). At the feed rate where the stable solution degrades to the feed temperature and zero conversion, PSR blowout (extinction) has occurred. The right endpoints of the curves in Fig. 1 represent the last stable solution just before blowout; hence, it is taken as the extinction point (approx. $F_{Ao} = 0.079 \text{mol} / s, T = 1530K, X_A = 0.738.$).

PSR AND LBV

Longwell and Weiss (1955) showed that, for an individual fuel/air pair, the experimental stirred combustor blowout feed rate passed through a maximum as φ was increased. Their observed dependence of blowout flow rates on φ is very reminiscent of the dependence of LBV on φ (Fristrom, 1995). In addition, Longwell and Weiss applied a global rate expression to these data to extract kinetic parameters.

In order to investigate this point further here, the Chemkin PSR code (Glarborg, Kee, Grcar, & Miller, 1986) was run with the Gas Research Institute C_1/C_2 detailed reaction mechanism (kinetics, species thermodynamics and heat capacities) version 2.11 (GRI, 1994). A hypothetical, adiabatic PSR of volume 250cm³ was used, with a feed temperature of 300K. Both the CH_4/air and H_2/air systems were considered.

The equations solved by the Chemkin PSR code are more generalized than Eq. (4) above. The species balances are

$$\dot{m}(Y_k - Y_k^*) - \dot{\omega}_k W_k V = 0,$$

(5a)

where m is the mass flow rate, Y_k the mass fraction of species k, * represents the feed, ω_k the net reaction rate of k (determined by Chemkin subroutine using the detailed mechanism), W_k the molecular weight of k, and V the PSR volume. The enthalpy balance is

$$\dot{m} \sum_1^K (Y_k h_k - Y_k^* h_k^*) + \dot{Q} = 0,$$

(5b)

Where h_k is the enthalpy of species k, Q the external heat loss rate, and K is the total number of species.

For each φ, the CH_4/air or H_2/air feed rate was incrementally increased until the PSR calculation converged to a blowout solution. Fig. 2 shows that these calculated PSR blowout mass flow rates maximize at approximately $\varphi=1.1$ for both CH_4/air and H_2/air. However, while the LBV values for CH_4/air maximize at$\varphi=1.1$, the maximum LBV for H_2/air occurs at approximately $\varphi=1.7$ (Fristrom, 1995). This apparent disparity is addressed in Section 5.

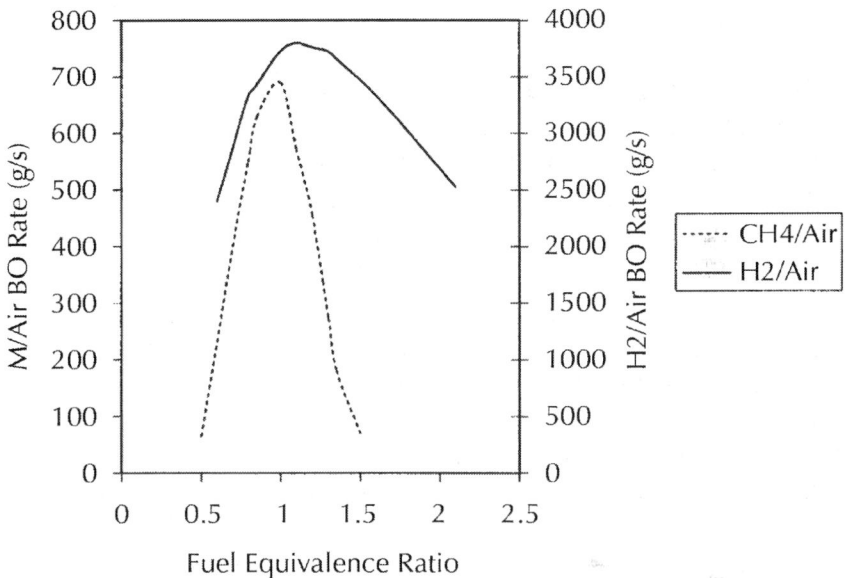

Figure 2: Hypothetical PSR blowout mass flow rates as functions of φ at 1 atm. For CH_4/air and H_2/air (feed temperature=300K).

Eq. (2b) shows that the LBV is a function of a chemical rate R and a thermal diffusivity. And show that, for the PSR, the mass rate at blowout is proportional to a chemical rate at blowout. Stirred reactor theory has been used to model turbulent premixed flame stabilization in the wake of a bluff body, such as a rod normal to the cold gas flow. Glassman (1987) shows that the unburned gas velocity U_{bo} at the point of extinction (blowout) of a rod (diameter d) stabilized flame is related to the LBV of the fuel/oxidant mixture:

$$S_L \propto \left[a \frac{U_{bo}}{d} \right]^{0.5}.$$

(6)

In this paper, then, the hypothesis is made that experimental LBV values can be correlated according to a modified form of Eq. (6):

$$S_L \propto \left[\frac{k}{\rho_u C_p} \frac{w_{bo}}{\rho V} \right]^{0.5} = \left[\frac{a}{\tau_{bo}} \right]^{0.5},$$

(7)

Where w_{bo} is the calculated PSR blowout mass rate The space time for the PSR at blowout, based on the high-temperature exit condition, is given by V/w_{bo}.

The thermal diffusivity and mass density will be calculated for the feed gas (fuel+oxidant) at 300K and the system pressure. The combustion of various fuel/oxidant pairs is treated with the PSR code and detailed reaction mechanisms taken from the literature. Calculated PSR blowout space times based in exit conditions are determined, and then correlated according to Eq. (7) with published LBV values. The adiabatic hypothetical PSR has a volume was 250cm³, with a 300K gaseous feed.

RESULTS AND DISCUSSION

Numerous fuel/oxidant systems were studied, as listed in Table 1. The PSR calculations containing C_1/C_2 hydrocarbon and nitrogen-based species were simulated with the GRI mechanism. For systems containing chlorinated species, the sub-mechanism of Ho, Barat, and Bozzelli (1992) was attached to the GRI listing. For C_{3+} hydrocarbon systems, the mechanism of Castaldi et al. (1996) was used.

Table 1: Data for Fig. 3

Fuel/oxidant	Pres. (atm)	$O_2/(O_2+other^a)$	Therm. diff. $a(cm^2/s)^b$	LBV (cm/s)	$\bar{\phi}^c$	LBV Ref.[d]	PSRboTemp. (K)	PSR $_{bo}$ (μS)	$(a/\tau_{bo})^{0.5}$ (cm/s)
CH$_4$/Air	0.1	0.21	2.20	57	1.0	7	1619	613	59.9
CH$_4$/Air	1.0	0.21	0.221	38	1.0	7	1725	80.9	52.3
CH$_4$/Air	5.8	0.21	0.038	20	1.0	7	1866	28.5	36.5
CH$_4$/Air	1.0	0.21	0.221	43	1.1	1	1775	86.7	50.5
CH$_4$/O$_2$/ N$_2$	1.0	0.98	0.228	326	0.98	5	2630	2.34	312
CH$_4$/O$_2$/ N$_2$	1.0	0.60	0.225	212	1.0	5	2346	5.16	209
CH$_4$/O$_2$/ He	1.0	0.21	1.19	130	1.06	7	1934	33.7	188
C$_2$H$_6$/Air	1.0	0.21	0.207	49	1.12	1	1771	76.2	52.1
C$_2$H$_4$/Air	1.0	0.21	0.208	78	1.1	1	1741	37.5	74.5
C$_2$H$_2$/Air	1.0	0.21	0.204	168	1.1	1	1722	15.3	115
C$_2$H$_2$/N$_2$O	1.0	NA^e	0.113	200	1.7	1	3092	3.84	172
C$_3$H$_8$/O$_2$	1.0	1.0	0.193	250	0.45	5	2127	3.00	254
C$_3$H$_8$/O$_2$	1.0	1.0	0.173	360	0.88	5	2597	2.14	284
C$_3$H$_8$/O$_2$	1.0	1.0	0.159	250	1.28	5	2641	3.20	223
C$_6$H$_6$/Air	1.0	0.21	0.20	47	1.08	1	1836	105	43.6

CH₃OH/Air	1.0	0.21	0.197	55	1.01	1	1621	57.8	58.3
CH₃Cl/CH₄Air	1.0	0.21	0.209	25	1.0ᶠ	2	1831	223	30.6
CH₂Cl₂/CH₄Air	1.0	0.21	0.210	18	1.03ᶠ	2	1339	139000	1.23
H₂/NO₂	1.0	NAᵉ	0.994	300	1.0	1	1837	8.66	339
H₂/Cl₂	1.0	NAᵉ	0.988	350	1.94	1	1886	10.0	314
H₂/Air	1.0	0.21	0.485	80	0.6	5	1152	27.3	133
H₂/Air	1.0	0.21	0.551	140	0.8	5	1178	18.2	174
H₂/Air	1.0	0.21	0.610	190	1.0	5	1189	15.2	200
H₂/Air	1.0	0.21	0.637	200	1.1	5	1182	14.6	209
H₂/Air	1.0	0.21	0.686	235	1.3	5	1181	14.2	220
H₂/Air	1.0	0.21	0.729	260	1.5	5	1182	14.5	224
H₂/Air	1.0	0.21	0.769	270	1.7	5	1166	15.3	224
H₂/Air	1.0	0.21	0.805	267	1.9	5	1147	16.6	220
H₂/Air	1.0	0.21	0.838	260	2.1	5	1134	18.2	215
H₂/O₂/N₂	1.0	0.50	1.21	450	3.0	5	1183	11.0	332
H₂/O₂/N₂	1.0	0.35	0.772	375	1.03	5	1234	6.10	356
H₂/O₂/N₂	1.0	0.985	1.14	890	1.18	5	1344	2.06	744
H₂/O₂/N₂	1.0	0.60	0.999	650	1.20	5	1323	3.15	563
H₂/O₂/N₂	1.0	0.60	1.17	600	2.15	5	1252	5.45	463

$H_2/O_2/CO_2$	1.0	0.50	0.98	500	1.7	5	1241	5.53	421
$H_2/O_2/CO_2$	1.0	0.80	0.992	700	0.99	5	1343	2.33	652
$H_2/O_2/CO_2$	1.0	0.90	1.20	800	1.69	5	1304	3.09	623
NH_3/O_2	1.0	1.0	0.193	114	0.85	1	2460	13.4	120
NH_3/N_2O	1.0	NA	0.13	72	0.95	1	2494	26.7	69.8
HCN/Air	1.0	0.21	0.195	55	1.13	6	1926	83.1	48.4
Town Gas[g]/Air	1.0	0.21	0.376	104	NA[e]	3	1724	71.3	72.6
CO/Air	1.0	0.21	0.220	45	1.7	1[h]	1296	306	26.8
$CO/O_2/H_2O$	1.0	0.99937	0.21	22	1.0	1	1800	404	22.8
$CO/O_2/H_2O$	1.0	0.8489	0.21	130	1.0	1	1504	28.0	86.6
HCN/O_2[i]	1.0	1.0	0.157	550	0.94	6	2691	3.03	228
CH_4/O_2[i]	21.4	1.0	0.0106	600	0.76	8	2780	0.135	280

N_2, He, H_2O, or CO_2

- Calculated at 300K for feed mixture; C_p and k for various species obtained from assorted handbooks.
- Fuel equivalence ratio.

- LBV References: 1, Fristrom (1995); 2, Gupta and Valeiras (1984); 3, Gunther (1974); 4, Jones (1993); 5, Lewis and von Elbe (1987); 6, Cohen and Simpson (1957); 7, Glassman (1987); 8, Strauss and Edse (1959).
- Not applicable.
- Molar ratio of fuels=1, φ incorporates total fuel.
- Composition (mole fraction): H_2 0.48, CO 0.09, CH4 0.29, CO2 0.04, N2 balance (Ref. 4).
- Fuel also contains 0.1 mol% H_2.
- Not plotted in Fig. 3.

A total of 46 individual cases were run in order to cover a wide range of fuels, oxidants, feed conditions, and LBVs. Fig. 3 presents the results from the systems listed in Table 1. All but the last two cases are plotted according to Eq. (7). There is an excellent linear dependence (R^2=0.9681). Inclusion of the last two cases from Table 1 statistically degrades the correlation, and is soon discussed.

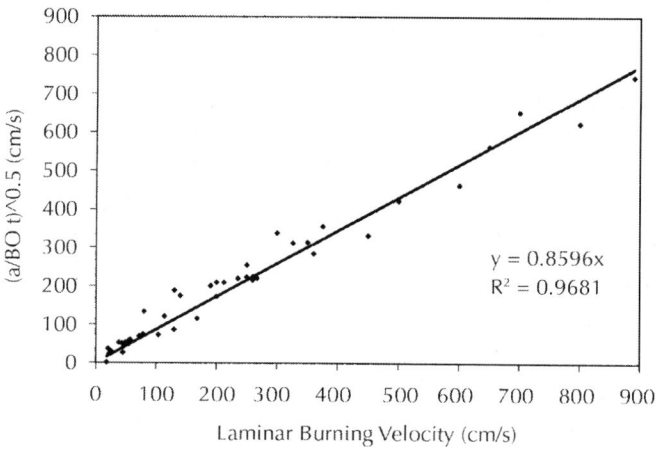

Figure 3: Correlation of literature laminar burning velocities with the square root of the ratio of feed thermal diffusivities to the space times at blowout of a hypothetical PSR.

The positive, linear correlation of the square root of the ratio of feed thermal diffusivity to hypothetical PSR space time at blowout with experimentally observed LBV accounts for the following:

- Wide range of LBV values (18-890cm/s).
- Wide range of feed thermal diffusivities (0.038–2.2).
- Range of hydrocarbon fuels (C_1, C2, C3, and C6).
- Oxyhydrocarbon fuel (CH_3OH).
- Fuel mixture (Town Gas).
- N-based fuels (NH_3, HCN).
- Range of diluents (CO_2, N_2, He).
- Range of fuel equivalence ratios (lean-stoichiometric-rich).
- Inhibition by chlorocarbons (CH_3Cl, CH_2Cl_2).
- Non-O_2 oxidants (NO_2, N_2O).
- Range of pressures (0.1-5.8atm).
- Range of fast H_2-based fuel systems.
- Slow CO-based systems.

Successful correlation with such a wide variety of systems suggests the validity of the correlation of Eq. (7).

A resolution of the apparent disparity between Fig. 2 for H_2/air and the corresponding literature LBV values is now apparent. Table 1 show that thermal diffusivities for H_2/air are about 3 times those for CH_4/air at a comparable   and pressure. Eq. (7) shows that the LBV depends on transport (a) and chemical kinetics (1/ $_{bo}$). However, transport rates are irrelevant in a PSR; only kinetics matter. The result is PSR blowout mass rates maximizing near =1.0 for either fuel (Fig. 2) The subset of cases in Table 1 for H_2/air (variation with) correlate well with Eq. (7) (Fig. 3) where both kinetics and transport are included.

Failure of the correlation for the HCN/O_2 and high pressure CH_4/O_2 cases requires comment. Fristrom (1995) recommends the LBV values of Cohen and Simpson (1957) for the HCN/air and HCN/O_2 systems. Curiously, the correlation easily handles the HCN/air case, but not HCN/O_2. Since the correlation correctly accounts for various levels of N_2 in a feed mixture as evidenced in several other cases, it is possible that a new experimental measurement of the LBV for

HCN/O_2 is needed. Using the Fig. 3 correlation, the $(a/ \tau_{bo})^{0.5}$ value of 228 for HCN/O_2 corresponds to a LBV of approximately 268cm/s, as compared to the 550cm/s value in Cohen and Simpson. The practical danger of working with HCN is the likely reason for the lack of more recent LBV data.

The relative success of the correlation for CH_4/air over a range of pressures (0.1-5.8atm) and for CH_4/O_2 and CH_4/He at 1atm suggests that its poor performance for the CH_4/O_2 case at 21.4atm could be due to one or more of the following: (1) The GRI mechanism is not accurate at 21.4atm. (2) The LBV data are not accurate or (3) the simple theory of Eq. (7) breaks down at high pressures. Further study is required.

CONCLUSIONS AND RECOMMENDATIONS

The observed laminar burning velocities of a wide range of fuel/oxidant systems have been well correlated with the square root of the ratio of the feed thermal diffusivity to the calculated space time (at exit conditions) at blowout of a hypothetical perfectly stirred reactor combusting the feed. The stirred combustor calculation, which reflects chemical kinetics alone, is not enough to correlate the burning velocities. Relative transport rates must be included, as with the thermal diffusivities. This is well represented by the H_2/air system. The method offers an engineering approach to estimating laminar burning velocities of fuel/oxidant pairs assuming detailed reaction mechanisms are available. It is recommended that this correlation be extended to even more fuel/oxidant systems, including fuel gas mixtures, elevated pressures, and various feed temperatures.

REFERENCES

1.   Barat, R. B. (1992). Jet-stirred combustor behavior near blowout: observations and implications. Combustion Science and Technology, 84, 187}197.

2.   Brouwer, J., Longwell, J. P., Saro"m, A., Barat, R. B., & Bozzelli, J. W. (1992). Chlorocarbon-induced incomplete combustion in a jet-stirred reactor. Combustion Science and Technology, 85, 87}100.

3.   Castaldi, M. J., Marinov, N. M., Melius, C. F., Huang, J., Senkan, S. M., Pitz, W. J., & Westbrook, C. K. (1996). Twenty sixth symposium (international) on combustion (pp. 693}702). Pittsburgh, PA: The Combustion Institute.

4.   Cohen, L. M., & Simpson, P. (1957). Burning velocities of HCN in air and oxygen. Combustion and Flame, 1, 60}62.

5.   Fogler, H. S. (1999). Elements of chemical reaction engineering (3rd ed.). Upper Saddle River, NJ: Prentice-Hall.

6.   Fristrom, R. M. (1995). Flame structure and processes. New York: Oxford University Press. Gas Research Institute * GRI, (1994). http://www.gri.org.

7.   Glarborg, P., Kee, R. J., Grcar, J. F., & Miller, J. A. (1986). PSR: a fortran program for modeling well-stirred reactors. Sandia Report SAND 86-8209. Livermore, CA: Sandia National Laboratories.

8.   Glassman, I. (1987). Combustion (2nd ed.). New York: Academic Press.

9.   Gunther, R. (1974). Verbrennung und Feuerungen. Berlin: Springer.

10. Gupta, A. K., & Valeiras, H. A. (1984). Burning velocities of chlorinated hydrocarbon}methane}air mixtures. Combustion and Flame, 55, 245}254.

11. Ho, W. P., Barat, R. B., & Bozzelli, J. W. (1992). Thermal reactions of CH Cl in H /O mixtures: implications for chlorine inhibition of CO conversion to CO . Combustion and Flame, 88, 265}295.

12. Jones, J. C. (1993). Combustion Science, Newtown NSW, Australia: Millennium Books.

13. Lewis, B., & von Elbe, G. (1987). Combustion, yames, and explosions of gases. New York: Academic Press.

14. Longwell, J. P., & Weiss, M. A. (1955). High temperature reaction rates in hydrocarbon combustion. Industrial and Engineering Chemistry, 47(8), 1634}1643.

15. Mao, F., & Barat, R. B. (1996). The interaction of fuel-bound nitrogen and fuel-bound chlorine during air-staged combustion. Combustion Science and Technology, 116}117, 339}357.

16. Nenniger, J. E., Kridiotis, A., Chomiak, J., Longwell, J. P., & Saro"m, A. F. (1984). Characterization of a toroidal well-stirred reactor. Twentieth symposium (international) on combustion (p. 473). Pittsburgh, PA: The Combustion Institute.

17. Strauss, W. A., & Edse, R. (1959). Burning velocity measurements by the constant-pressure bomb method. Seventh symposium

(international) on combustion (p. 377). Pittsburgh, PA: The Combustion Institute.

18. Westbrook, C. K. (1983). Numerical modeling of #ame inhibition by CF3Br. Combustion Science and Technology, 34, 201}225.

An Inward and Outward Natural Gas Hydrates Growth Shell Model Considering Intrinsic Kinetics, Mass and Heat Transfer

Bo-Hui Shi[a], Jing Gong[a], Chang-Yu Sun[b], Jian-Kui Zhao[a, 1], Yao Ding[a], and Guang-Jin Chen[b]

[a]Beijing Key Laboratory of Urban Oil and Gas Distribution Technology, Department of Oil & Gas Storage and Transportation Engineering, China University of Petroleum, Beijing 102249, China
[b]State Key Laboratory of Heavy Oil Processing, China University of Petroleum, Beijing 102249, China

ABSTRACT

The natural gas hydrates formation and growth at 2 MPa and 277.15 K were studied at different water cuts for water-in-condensate oil emulsions in the flow loop unit. The variations of gas consumption with time at different water cuts were obtained. The experimental results showed that the gas consumption value increased with the rise of water cut. The total water conversion rates not only depended on the water cuts, but also related to several other factors, two of which were the amount of the dissolved gas and the surface/volume ratio of the particles. No more natural gas transformed into hydrates after 3 h, which was likely caused by intrinsic kinetics, mass transfer and heat transfer limitations. An inward and outward natural gas hydrates growth shell model was proposed considering all the three limitations to simulate the gas and water consumptions to form natural gas hydrates.

INTRODUCTION

Natural gas hydrates have been known to plug transportation pipeline in oil and natural gas industry [1]. They are crystalline solids composed of water and gas molecules. The gas molecules (guests) are trapped in water cavities (host) that are composed of hydrogen-bonded water molecules [2]. In recent years, gas and oil industry has moved towards deep-water exploration and production. The condition of the subsea is ideal for the formation of clathrate hydrates of natural gas, which can potentially block the subsea pipelines. It could be a major problem in flow assurance for offshore petroleum industry. Traditionally, the subsea pipelines are designed to operate outside the hydrates formation region by adding inhibitors or insulation. While these techniques have been proved to reach their limits, new remediation techniques have been presented like adding anti-agglomerate low dosage additives or cold flow, in which hydrate particles are evacuated with the liquid flow as a slurry to prevent natural gas hydrates plug in petroleum production pipelines [3] and [4].

Hydrates formation and growth mechanism are key issues to apply the pseudo-slurry flow technology for the subsea pipeline transportation in offshore fields. Thermodynamic models of hydrates formation have been proposed and established extensively [5] and [6]. Meanwhile, lots of researchers have developed various hydrate kinetics models at different experimental conditions [7], [8], [9] and [10]. And mass and heat transfer models for the lateral hydrate-film growth along water/guest-fluid interfaces have been experimentally and theoretically studied by several research groups [11], [12], [13], [14], [15], [16], [17],[18] and [19]. All the theoretical and experimental studies were carried out to reveal that hydrates formation and growth mechanism were inherently complex and system dependent, involving multiphase flow and possible limitations by intrinsic kinetics, mass transfer and heat transfer [20]. It is too difficult to establish a hydrates growth model considering all these aspects, especially during the pipeline flow with multicomponent fluids in various multiphase flow regimes [19]. Jamaluddin et al. [21] developed a mathematical model to simulate the hydrates formed on the free surface of gas-water contact, which coupled intrinsic hydrate formation kinetics with heat and mass transfer phenomena. But a wide variety of evidences in literatures strongly suggested that natural gas hydrates would form on the dispersed water droplets in multiphase transport pipelines [12], [19], [22] and [23]. Gong et al. [23] combined two limitations of kinetics and mass transfer to propose a hydrates shell model to numerical simulate the growth of the hydrates during flow condition, and ignored the heat transfer limitation.

In this work, the variations of natural gas consumption for hydrates formation with time at different water cuts (10%, 20%, and 25%) were measured for the water-in-condensate oil emulsion with anti-agglomerate additives in the flow loop unit. A new shell model was proposed for natural gas hydrates growth, in which three limitations of intrinsic kinetics, mass transfer and heat transfer were considered, followed by both inward and outward hydrates growth around the water droplet.

EXPERIMENTS

Experimental Equipment

The main body of the flow system (Fig. 1) was a transparent U-bend double pipe of 20 m in length (made by DB Robinson Corp. Canada) and 25.4 mm in inner diameter, which could sustain a system pressure up to 4 MPa [24]. Coolant (ethylene glycol solution) was circulated in the annulus of the double pipe through two thermostats (Neslab RTE 111D) to maintain the temperature at a range of −25 °C to 50 °C.

Six thermocouples calibrated to ±0.1 K were installed along the flow loop. Precision Heise pressure gauge (accurate to ±0.1%, scale from 0 to 4.0 MPa) was used to measure the system pressure. The pipe was connected to a refrigerated mixing tank. A centrifugal pump was used to circulate the heterogeneous mixture through the pipe loop. A laser granulometer (made by OMEC Corp.) was installed in the pipe loop for measuring the hydrate particle size distribution and solid particle concentration. Gas consumption was measured on line by electronic balance (made by Sartorious Corp. Germany).

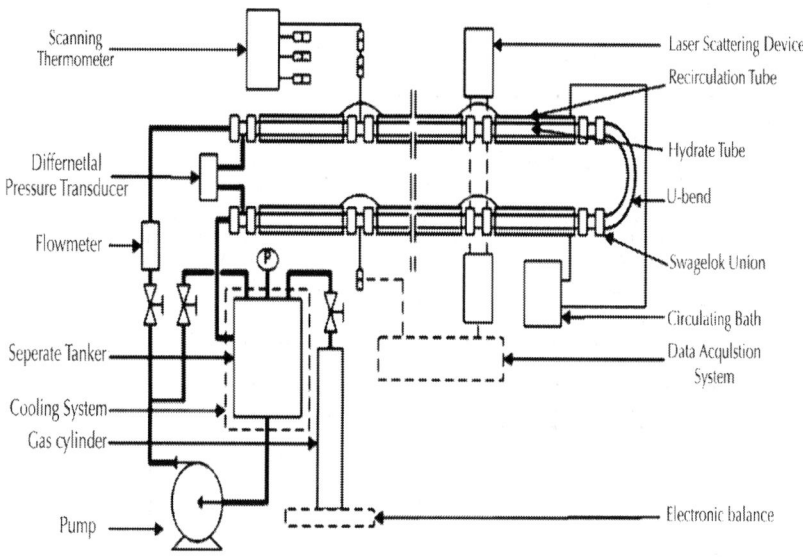

Figure 1: Schematic of the flow loop system.

Experimental Procedure

Synthetic natural gas was supplied by Beijing Beifeng Gas Industry Corporation, with the composition of 82.18% CH_4, 13.89% C_2H_6, and 3.93% C_3H_8. The condensate oil was supplied as dead with composition as listed in Table 1, which was sampled from JZ well 2# offshore gas field located in Bohai Sea of China, with the density of 0.75 g/cm³ (293.15K, 0.1 MPa) and viscosity of 1 cp. The ratio of gas phase to liquid phase was 7:8. Fresh water taken from the domestic network was used as water phase. The anti-agglomerant (made by China University of Petroleum, Beijing) was used to prevent the hydrate crystals from aggregating, so that the hydrates formed in the water-in-condensate oil emulsion could keep flowing without any blockages. The anti-agglomerant was a mixture of several types of chemical additives. Its performance has been verified for water + light oil + natural gas flow system [25].

Table 1: Composition of the condensate oil

Component		Mol (%)
C6	Hexanes	2.98
C7	Heptanes	3.90
C8	Octanes	8.55
C9	Nonanes	7.07
C10	Decanes	6.46
C11	Undecanes	5.39
C12	Dodecanes	5.64
C13	Tridecanes	6.91
C14+	Tetradecanes plus	53.10

The experimental procedure was described as follows:

- Condensate oil and water systems were prepared with different water cuts. The total volume of liquid phase was 16 L.

- The flow loop was cleaned with detergent solution and pure water successively. After it was vacuumized, the water-in-condensate oil emulsion with certain dosage of anti-agglomerant was injected into the mixing tank.

- The refrigerating system was turned on and cooled to desired temperature. In this work, the experimental temperature was kept at 277.15 K.

- After the specified temperature was reached, synthetic natural gas was slowly injected into the mixing tank until the pressure of the flow loop system was 2 MPa. Then the pump was turned on and the fluid was circulated in the flow loop. Meanwhile, the bulk pressure in the loop was kept constant via the supplement of gas from a gas cylinder. The amount of gas consumed was weighted by an electronic balance on line. It was assumed that the liquid could be saturated with the natural gas quickly in the flow condition. The variation of gas consumption with time was collected since the start-up of the circulation pump.

- The formation and growth of natural gas hydrates were tested for nearly 30 h. Afterward, the pump was turned down and hydrates dissociated in the pipeline due to the temperature rise and pressure drop.

For other water cut, the above steps were repeated. The mean flow rate was kept constant at given water cut, but increased by increasing water cuts. The flow velocities at water cuts of 10%, 20%, and 25% were 0.2175 m/s, 0.3054 m/s, and 0.3266 m/s, respectively.

RESULTS AND DISCUSSION

Gas Consumption

During the experiments, the gas consumption for hydrates formation was monitored by the electronic balance. The total gas consumption as function of time at different water cuts (10%, 20%, and 25%) were shown in Fig. 2, which included both the amount of gas consumed by hydrates formation and the amount of gas dissolved into the liquid hydrocarbon phase. During the injection of natural gas, the circulation pump was shut down and it was difficult for gas to dissolve in the liquid phase under static condition. When the pump was started up, the hydrocarbon liquid approached its gas saturation level relatively quickly and it was assumed that this process had nearly come to an end before the hydrates formation commenced, though the two processes appeared partly overlapping at the end of the saturation stage. The amount of gas that would be dissolved in the hydrocarbon liquid at the saturation level was given by the system pressure and temperature and could be calculated by the thermodynamic flash calculation [26]. The onset of hydrates formation was defined as the time where the total gas consumption just exceeded the amount of gas that could be dissolved in the hydrocarbon liquid at saturation level. According to this definition and from the gas consumption data as shown in Fig. 2, it was estimated that the onset time of hydrates formation during experiments with 10%, 20% and 25% water cuts was 8.13, 2.80 and 2.45 min after pump was started up respectively. The total amount of synthetic natural gas dissolved at the saturation level was a function of oil phase volume and was thus lower at higher water cut. Meanwhile, the corresponding mean flow rate was larger at higher water cut in our experiments. Therefore, the time for complete liquid phase saturation by gas was shorter at higher water cut.

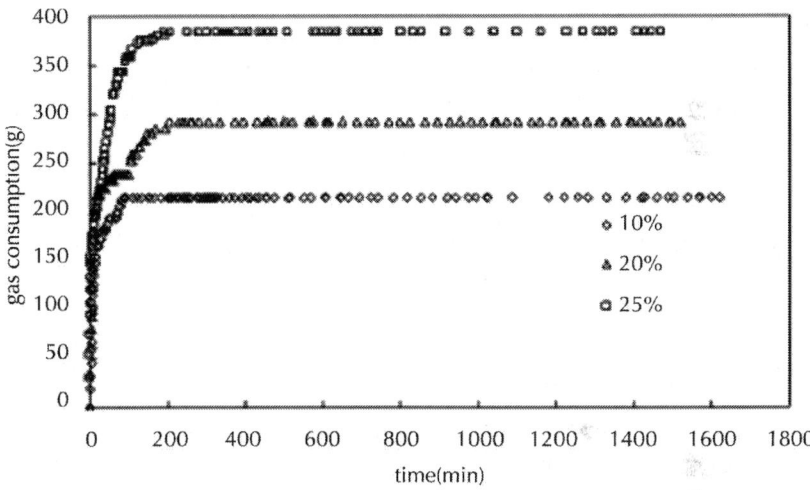

Figure 2: Variations of natural gas consumptions at different water cuts with time during experiments (2 MPa, 277.15 K).

After the gas dissolving process, the gas consumption was assumed as resulting from the hydrates formation. As shown in Fig. 2, the hydrates formation processes were completed within the first 3 h of the experiments and from that time, the gas consumption rates were close to zero throughout the remaining part of the experimental runs. The amount of gas consumed increased with the accretion of water cut from the curve analysis in Fig. 2. The reason for this phenomenon was that both of the total water surface and the degree of the water droplet dispersion were enhanced at higher water cut, which were consistent with the increase in initial gas consumption rates [22].

Water Conversion Rate

A two-step hydrates formation mechanism proposed by Chen and Guo [5] was used to predict hydrates formation condition, gas composition, and the fraction of the linked cavities occupied by the gas molecules in the hydrates. It was known that the gas composition in the vapor phase would change with the formation of hydrates. For the experimental system used in this work, gas consumed by hydrates formation was compensated continuously by fresh gas supplement to maintain the constant system pressure. Therefore, it was regarded as semi-closed experimental system. In view of the continuous gas consumption and injection of fresh gas, the gas composition in the flow loop would fluctuate during the experimental process. The stable gas composition could not be obtained at the dynamics conditions. Therefore, we did not consider the variation of gas composition during the experimental procedure. Only the compositions of initial fresh gas and the residual gas after the shut down of pump were analyzed by a gas chromatograph. The fluctuation of gas composition during the hydrates formation process was then deduced from the compositions of fresh gas and residual gas. During the following simulation work, the equilibrium hydrate formation pressure was assumed as a linear distribution function with time in the region between 1.1 MPa and 1.7 MPa at 277.15 K, due to the fluctuating gas composition.

In general, mainly structure II hydrate was expected to form with natural gas under the special conditions [27]. The theoretical numbers of water molecules per gas molecule of structures I and II were 5.75 and 5.67, respectively [28]. The real hydration number at this experimental condition could be estimated by a correlation function of the fraction of the linked cavities occupied by the gas molecules in hydrates mentioned in Ref. [29]. According to the structure characteristics of hydrates and the property of the synthetic natural gas [30], the water conversion rates at different water cuts could be estimated from the gas consumption for hydrates formation, which was listed in Table 2. It was seen that the total mass of gas consumed at the end of process increased with the accretion of water cut, and the relationship between the total water conversion rates and the different water cuts was not clear. The water conversion rate depended on several other factors, two of which were the amount of the dissolved gas and the surface/volume ratio of the particles. It was known that the amount of the dissolved gas decreased with the increasing of the water cut, while the surface/volume ratio of the particles was inversely proportional to the initial radius of the water droplet, which was also a vital parameter. Sun [31] adopted the same flow system to study the size distribution of the hydrate particles using a laser granulometer. It was found that the average diameter of hydrate particles mainly depended on the flow rate. The initial mean radiuses of the water droplet used in our work were predicted using a correlation model developed by Turner et al. [20], and the results were listed in Table 3. This correlation model [20] was a function of the fluid properties, interfacial tension, and the flow rate, which was validated by the experimental data of water-in-crude oil emulsion systems. From Table 3, it could be seen that the particle sizes decreased with the increasing of the flow rate, while the flow rates increased with increasing of the water cut for our experiments. Therefore, the surface/volume ratio of the particle increased with the rise of the water cut as shown in Table 3. Based on the opposite influence of the dissolved gas amount and the surface/volume ratio of the particles, the relationship between the total water conversion rates and the water cuts was irregular.

Table 2: Total water conversion rates at the different water cuts

Water Cut (%)	Total water mass (g)	Total gas consumption mass (dissolved) (g)	Rcon estimated (%)
10%	1600	214 (113.35)	44.67
20%	3200	291(101.66)	42.01
25%	4000	383 (95.81)	50.98

Table 3: Sizes of water droplet at the different water cuts

Water cut (%)	Mean velocity (m/s)	Shear rate (s⁻¹)[a]	Flow index[b]	Initial mean radius of water droplet (μm)	Surface/ volume ratio (1/m)
10%	0.2175	74.83	0.7303	86.56	11553
20%	0.3054	112.94	0.5893	54.57	18325
25%	0.3266	131.82	0.4703	46.19	21650

Shear rate is given by the Rabinowitsch–Mooney equations for power law fluid.

bAs determined form the analysis of Zhao [32].

During the experimental period, all the water conversion rates at different water cuts did not reach 100% at the constant system pressure. This phenomenon supported a natural gas hydrates growth shell model limited by intrinsic kinetics, mass transfer and heat transfer. The unconverted water was trapped inside the formed hydrate shell particles.

MODELING

The precondition to natural gas hydrates formation and growth continuously included the required thermodynamic conditions with enough driving force of the kinetic formation, continuous mass transfer of gas and water, and rapid heat transfer of exothermic hydrates formation. Based on the necessary conditions mentioned above, a new natural gas hydrates growth shell model considering kinetics, mass and heat transfer was developed, with inward and outward growth of the hydrate shell. The schematic diagram of natural gas hydrates growth shell model was shown in Fig. 3.

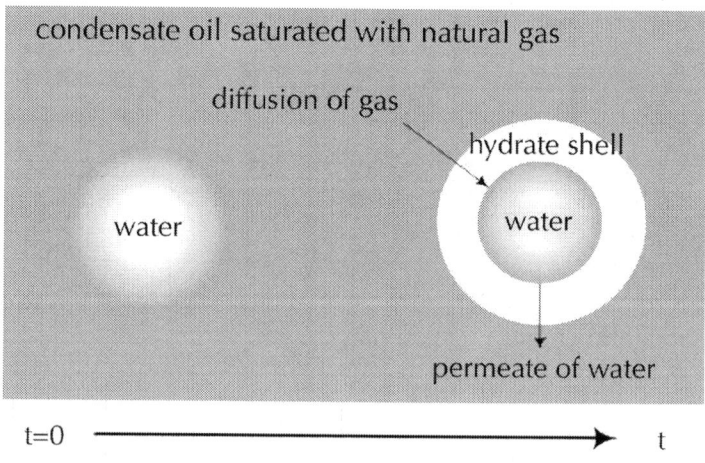

Figure 3: Schematic diagram of hydrates growth shell model.

Lee et al. [33] found that nucleation occurred at a single or multiple sites and was followed by the lateral growth of the hydrate film covering the surface of the droplet within 25 s. Therefore, we assumed that the initial hydrate film covered the water droplet quickly, and the induction time of hydrates formation was ignored in this work. After the droplet was covered with hydrate film, further hydrates growth depended on the diffusive transfer of gas to the inner surface of the hydrate shell and the capillary driven flow of water across the hydrate shell to the outer side of the hydrate shell [33] and [34]. Meanwhile, the process of hydrates formation was an exothermic process. The heat removal by conductive heat transfer was considered to be balanced with the released heat of hydrates formation [12], [15] and [16]. The temperature profile around the water droplet during the growth process could be estimated in our model. The concentration of natural gas in the liquid phase depended on the temperature around the hydrate shell. The driving force of the kinetics formation was the difference between the gas concentration at the growth interface and that at the three-phase equilibrium condition. So, the interactive effects of kinetics, mass transfer and heat transfer on hydrate growth on water droplets could be established.

In our model, it was assumed that the inward hydrates growth took place at the hydrate/water interface (H/W), and the outward hydrates growth took place at the hydrate/oil interface (H/O) as shown in Fig. 3. Actually, the real process of hydrates growth should occur at the inner and outer surfaces of the hydrate film simultaneously. However, for simplifying the hydrates growth shell model, we calculated the water consumption for hydrates inward growth first with a fixed outer diameter of the hydrate shell. Then we calculated the water consumption for hydrates outward growth with the calculated inner diameter of the hydrate shell. Meanwhile, we assumed that the water droplet and the hydrate shell were kept in a sphere shape. It was known that the volume of consumed water was expanded by 1.25 times during its conversion into hydrates [35]. Our estimates of the final inner and outer radiuses of the hydrate shell particles at each time step during the process should be corrected by use of this coefficient.

Inward Hydrates Growth at *H/W* Interface

Gas Diffusion through the Hydrate Shell to H/W Interface

The liquid phase and the gas phase could mix very well in the flow condition and be saturated by gas quickly. The synthetic natural gas was a mixture of many components. Based on the thermodynamic model calculation, the concentrations of each component in the liquid phase could be estimated. We assumed that the concentrations of each component in the oil and those at the *H/O* interface were equal at the beginning of the simulation. The concentrations of each component in the liquid were dependent on the temperature and pressure around the hydrate shell, while the concentrations of each component at *H/Winterface* depended on the diffusion flux of each gas hydrate former across the hydrate film.

In general, the hydrate formers of natural gas include CH_4, C_2H_6, C_3H_8, i-C_4H_{10}, n-C_4H_{10}, N_2, H_2S and CO_2[2]. In our work, the number of gas hydrate formers was three, including CH_4, C_2H_6, and C_3H_8. The gas diffusion rate of each gas hydrate former component j at *H/W* interface could be extended from that established by Gong et al. [23]. The total gas molecule diffusion rate was equal to the sum of the diffusion rate of each component, which was expressed as

$$W_{d.H/W} = \sum_{j=1}^{3} \left[-4\pi D_{f,j} \frac{(C_{H/O,j} - C_{H/W,j})}{(1/r_i^{\Delta t-1} - 1/r_o^{\Delta t-1})} \right],$$

$$(1)$$

where $D_{f,j}$ was the diffusivity of component j, $C_{H/W,j}$ and $C_{H/O,j}$ were the concentrations of component j at*H/W* interface and *H/O* interface, $r_i^{\Delta t-1}$ and $r_o^{\Delta t-1}$ were the inner and outer radiuses of the hydrate shell at the end of the last time step $\Delta t-1$, and $W_{d,H/W}$ was the total gas molecule diffusion rate at *H/Winterface*.

Gas and Water Consumed For Hydrates Formation at the H/W Interface

At equilibrium condition, the fugacity of a hydrate forming gas component in the vapour phase must be equal to the fugacity of the same component in the liquid phase. Thus the equilibrium hydrate formation properties of the hydrocarbon fluid could be described by the vapour composition only through the two-phase gas–liquid hydrocarbon region. The gas consumption for hydrates formation at the H/W interface depended on the kinetics driving force of the difference between the concentration of the gas former at the system and that at the equilibrium condition. From the theory of Englezos et al. [7] and [8], the Henry constant was a crucial parameter to determine the gas consumption rate. However, it was difficult to define the Henry constant of each gas component in the condensate oil system in this work. Therefore, we defined a concentration parameter to describe the degree of the gas dissolution, which was the ratio of the fugacity to the concentration ($\Omega = \varphi P/Coil$). A modified and simplified kinetic model from that developed by Gong et al. [23] was established for the mixture as follows:

$$W_{g,H/W} = \sum_{j=1}^{3} \left[4\pi \left(r_i^{\Delta t-1} \right)^2 K_j^* \left(\Omega_j C_{H/W,j} - \Omega_{eq,j} C_{eq,j} \right) \right],$$

(2)

where $C_{eq,j}$ was the concentration of component j at the three-phase equilibrium pressure; Ω_j was the concentration parameter of component j at the system condition; $\Omega_{eq,j}$ was the concentration parameter of component j at the three-phase equilibrium pressure, and $W_{g,H/W}$ was total gas consumption rate at the H/W interface. Since the diffusion limitation of the inward hydrates growth was considered in the diffusion model, K_j^* only represented the kinetic rate constant of component j in the adsorption process.

Water consumption rate was in proportion to the radius of water droplet. The water consumption rate at H/W interface was the same as that proposed by Gong et al. [23].

Mass Conservation during the Inward Hydrates Growth at H/W Interface

In the quasi-steady condition, it was assumed that the rate of gas diffusion through the hydrate shell to the H/W interface was instantaneously balanced with the rate of hydrates formation controlled by the intrinsic kinetic model. Meanwhile, the consumption rate of gas molecules was in proportion to the water consumed during hydrates formation and growth [23]. Therefore, the mass conservation equation at H/W interface was established for the variation of inner hydrate shell radius with time as follow:

$$-\frac{dr_i}{dt} = \frac{\beta M_w}{\rho_w} \sum_{j=1}^{3} \left[\frac{\Omega_j C_{H/O,j} - \Omega_{eq,j} C_{eq,j}}{1/K_j^* + \Omega_j \left(r_i^{\Delta t-1}\right)^2 \left(1/r_i^{\Delta t-1} - 1/r_o^{\Delta t-1}\right)/D_{f,j}} \right],$$

(3)

where β was the hydration number, ρ_w was the density of water, Mw was the molecular weight of water, and r_i was the inner radius of the hydrate shell and t was the time. Eq. (3) was nonlinear in terms of the radius of the inner surface of the hydrate shell, and was required iteration to obtain an accurate solution. The common fourth-order Runge–Kutta method was used to obtain approximate value of the ordinary differential equation. The concentration of gas at the H/W interface could also be calculated by the mass conservation during the inward growth. The total consumption of water during the calculation step Δt at the H/W interface, Vw,H_iW, was given by:

$$V_{w,H/W} = \frac{4}{3}\pi \left[\left(r_i^{\Delta t-1}\right)^3 - \left(r_i^{\Delta t*}\right)^3 \right],$$

(4)

where $r_i^{\Delta t*}$ was the radius of the inner surface of the hydrate shell at the end of the calculated time step Δt only considering the inward hydrate growth, which was an intermediate variable and was not the final inner radius of the hydrate shell.

Outward Hydrates Growth at *H/O* Interface

We assumed that the water permeated from the core of the water droplet was instantaneously converted into hydrates, because the gas concentration at H/O interface was sufficient for hydrates formation and growth. The intrinsic kinetic control factor was ignored at H/O interface for the hydrates formation. All the permeated water from the core droplet was totally converted into gas hydrates. The degree of the hydrates formation and growth at H/O interface depended on the amount of water permeated. This made the outside of the hydrate shell dry and inert. Meanwhile, with the addition of anti-agglomerant in our experiments, the hydrate shell particle was not easy to agglomerate with other hydrate shell particles.

In our work, the assumptions to describe the permeation of water were referred to the model developed by Mori and Mochizuki [11] with incorporating some alterations. The thickness of the hydrate shell was uniform and was defined by the difference between the outer radius of the hydrate shell at the last time step and the inner radius of the hydrate shell calculated by the inward hydrates shell model. Micro perforations and tortuous capillaries with the same radius and length were evenly distributed over the shell. The driving force for sucking the water was the capillary pressure induced by the water and oil interface, which was located near the oil-side mouth of each capillary and was strongly concave towards the oil side due to the hydrophilic nature of the hydrate surface [36]. The total volume flow rate though the hydrate shell was estimated by a Hagen–Poiseuille flow driven by the capillary pressure, which was given by:

$$q_{water} = n_c \frac{r_c^3 \cos\theta}{\tau} \frac{\pi\sigma}{4\mu_w(r_o^{\Delta t-1} - r_i^{\Delta t*})},$$
(5)

where n_c was the number density of capillaries, r_c was the radius of the capillaries, τ (≥ 1) denoted the tortuosity of the capillaries, θ was the water-side contact angle on the capillary wall, σ was the water/condensate oil interfacial tension, and μ_w was the viscosity of water. None of these capillary parameters could be measured easily. For simplification, we defined a porous parameter to describe the property of the hydrate shell ($\varepsilon H = n_c r_c^3 \cos\theta/\tau$). So the total volume of water permeated in the calculated time step Δt, Vw,H_2O, was given by:

$$V_{w,H/O} = \frac{\pi\varepsilon_H\sigma\Delta t}{4\mu_w(r_o^{\Delta t-1} - r_i^{\Delta t*})}$$
(6)

Temperature Profile around the Water Droplet

The heat of hydrates formation was referred to Sloan and Koh [2]. The total heat released during hydrates formation in the calculated time step Δt could be defined as:

$$Q_{heat}^{\Delta t} = \Delta H \frac{\rho_w}{\beta M_w}(V_{w,H/W} + V_{w,H/O})$$
(7)

In the quasi-steady condition, the energy conservation equation was given as [18], [21] and [37]:

$$0 = \lambda \left[\frac{1}{r^2} \frac{d}{dr} \left(r^2 \frac{dT}{dr} \right) \right],$$

(8)

where λ was the thermal conductivity, r was the radius of water droplet covered by hydrate shell and T was the temperature. The temperature of the system at infinity was kept at 277.15 K by the thermostats.

It was assumed that the heat generated by hydrates formation instantaneously diffused away into the water droplet and the condensate oil, and would not accumulate in the hydrate shell [18]. Fourier's law was applied to model the heat transfer through the hydrate shell. The boundary conditions at both sides of the hydrate shell were given as follows:

$$\frac{\lambda_w}{\lambda_w + \lambda_{oil}} \frac{Q_{heat}^{\Delta t}}{4\pi \Delta t \left(r_i^{\Delta t} \right)^2} = -\lambda_w \left. \frac{dT}{dr} \right|_{r=r_i^{\Delta t}}$$

(9a)

$$\frac{\lambda_{oil}}{\lambda_w + \lambda_{oil}} \frac{Q_{heat}^{\Delta t}}{4\pi \Delta t \left(r_o^{\Delta t} \right)^2} = -\lambda_{oil} \left. \frac{dT}{dr} \right|_{r=r_o^{\Delta t}},$$

(9b)

where λ_w and λ_{oil} were the thermal conductivities of water and condensate oil. The concentrations of each gas component at H/O interface depended on the temperature around the hydrate shell. The temperature could be estimated by combining Eqs. (8), (9a) and (9b). Thus, the heat transfer limitation around the hydrate shell for hydrates formation and growth could combine with the gas diffusion model mentioned above.

Implementing the Inward and Outward Hydrates Growth Model

With numerical analysis of the equations mentioned above, we could simulate the variations of the radiuses on both sides of hydrate shell and the water conversion rates during the calculated time interval caused by water consumption for inward and outward hydrates formation and growth. The gas consumption for hydrates formation could be given by the total volume of water consumption. In addition, the temperature profile around the hydrate shell was estimated by the heat transfer model.

In fact, it was difficult to determine the kinetics and diffusivity of each different gas hydrate former[7] and [8]. For simplification, we assumed the kinetics and diffusivity of the three components of natural gases were identical in this work. Since the individually determined kinetic constant rate was dissimilar under different experimental conditions [19], the kinetic constant rate regressed from the data was 6.0×10^{-4} mol/(m² MPa s) in this work, which was different from that estimated value by Englezos et al.[7] and [8]. The flow rate impact on kinetics was adjusted for fitting the model parameters, which would be defined in the future for model development.

With the hydrates formation and growth, the thickness of the hydrate shell became larger and the structure of hydrate shell was more compact. The channel for gas diffusion and water permeation would be more complex and the efficiency of the mass transfer would descend with the hydrates growth. Therefore, the diffusivity of the gas and the porous property parameter of the hydrate shell would not remain constant during the hydrates growth. There would be a relationship between these two parameters and the hydrates formation volume fraction. We selected the logarithm mathematics to describe these relationships, which were expressed by the following equations:

$$D_f = -D_{f,0} \; \ln(-\xi\psi), \; \psi > 0$$

(10a)

$$\varepsilon H = -\varepsilon_{H,0} \ln(-\zeta\psi), \psi > 0$$

(10b)

where $D_{f,0}$ and $\varepsilon_{H,0}$ were the initial diffusivity of the gas and the initial porous parameter of the hydrate shell with $\psi = 0$; ξ and ζ were the mass transfer efficiency parameters adjusted by the experimental data. If the values of the diffusivity and the porous parameters calculated by Eq. (10a) and (10b) were lower than zero, we defined that their values were equaled to zero.

Fig. 4(a–c) showed the calculated and experimental total gas consumption during the experiments at the three water cuts as function of time. The left side of Fig. 4(a–c) showed the calculated and experimental total gas consumption during the experiments. The right side of Fig. 4(a–c) focused on the first 3 h of the experiments. It was seen that no more natural gas transformed into hydrates after 3 h. The simulation results were predicted by simultaneous numerical iteration of the calculation procedure. For all the experiments, the hydrates growth shell model fitted the experimental data well. All the mass transfer parameters used to simulate the hydrates formation and growth processes were given in Table 4. The mass transfer parameters of the model for all experiments were different from each other of the scenarios. The range of the diffusivity was 1.59×10^{-12} to 6.45×10^{-12} m^2 s. The regressed diffusivity magnitude was consistent with a previous estimation of 5×10^{-14} to 5×10^{-10} m^2 s presented by Makogon [38]. It was found that all the initial porous parameters decreased with the increment of the water cut. The reason was that the initial mean radius of water droplet decreased with the accretion of the water cut, which was caused by the flow rate. The larger water droplet was easier for mass transfer. The results indicated that the water conversion was strongly dependent on the average radius of the water droplet in the condensate oil. According to the model proposed by Turner et al. [20] for defining the initial water droplet, the properties of the fluid and the flow condition were critical to determining the mass transfer parameters. Due to the variation of the vapor gas phase composition in this experimental system, the actual final total water conversion rates given in Table 4 were larger than the values predicted based on the hydrate structure characteristic and the property of the synthetic natural gas listed in Table 2.

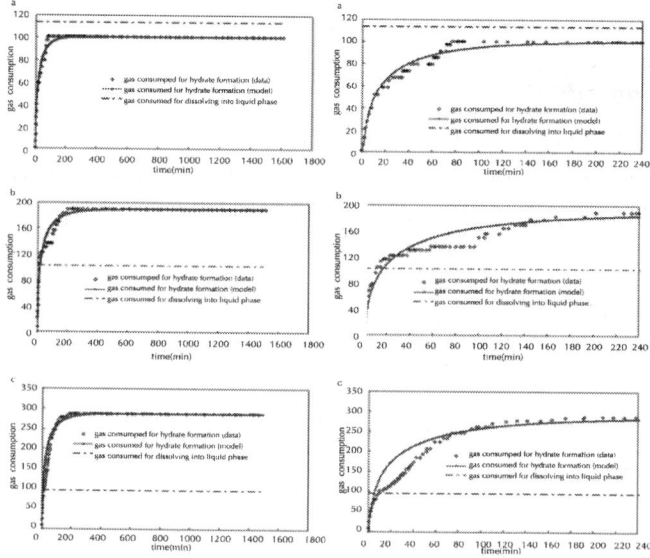

Figure 4: Gas consumption with time during hydrates formation in the flow loop during experiments at (a) 10% water cut, (b) 20% water cut, (c) 25% water cut (2 MPa, 277.15 K).

Table 4: Modeling parameters and calculated results corresponding to the illustrations reported in Fig. 4(a–c) and Fig. 5(a–c)

Water cut (%)	D0 (m² s)	ξ	$\varepsilon H,0$ (m³)	ζ	ri(µm)	ro(µm)	AAD-1ª(%)	AAD-2ª(%)	Rcon calculated (%)
10%	6.45 × 10⁻¹²	1.830	5.31 × 10⁻²⁷	0.852	69.11	89.96	5.2209	0.7956	49.095
20%	1.18 × 10⁻¹²	1.932	6.76 × 10⁻²⁹	2.124	44.38	56.59	8.4418	1.7984	46.185
25%	1.59 × 10⁻¹²	1.628	5.94 × 10⁻²⁹	0.001	35.12	48.25	10.934	1.8099	56.031

$$^a\ \mathrm{AAD}(\%) = \frac{1}{N} \sum_{j=1}^{N} \left[\frac{|m_{cal.} - m_{exp.}|}{m_{exp.}} \right]_j \times 100.$$

The left side of Fig. 5(a–c) showed the calculated inner and outer radiuses of the hydrate shell during the experiments at the three water cuts as function of time. The right side of Fig. 5(a–c) focused on the first 3 h of the experiments. The specific data of the inner and outer radiuses at different water cuts at the end of the process were also listed in Table 4. The shell growth model predicted that the unconverted water was trapped inside the formed hydrate shell particles. According to Fig. 5(a–c) and Table 4, the growth stopped when the shell thickness reached some limit (20.85 µm for the larger particles at 10% water cut and 12.21 and 13.13 µm for the smaller particles at 20 and 25% water cut). The shell thickness at 20% water cut was lower than that at 25% water cut, which was consistent with the trend of total water conversion rate from 20 to 25% water cut. The results indicated that the inward growth rate at the *H/W* interface was faster than the outward growth at the *H/O* interface. It could be explained by the model assumption. The further growth at both sides of the hydrate shell was strongly dependent on the mass transfer of the gas diffusion and the water permeation. The gas diffusion across the hydrate shell to the *H/W* interface was much easier than water permeation, because the size of the gas molecule was fairly smaller than that of the water molecule.

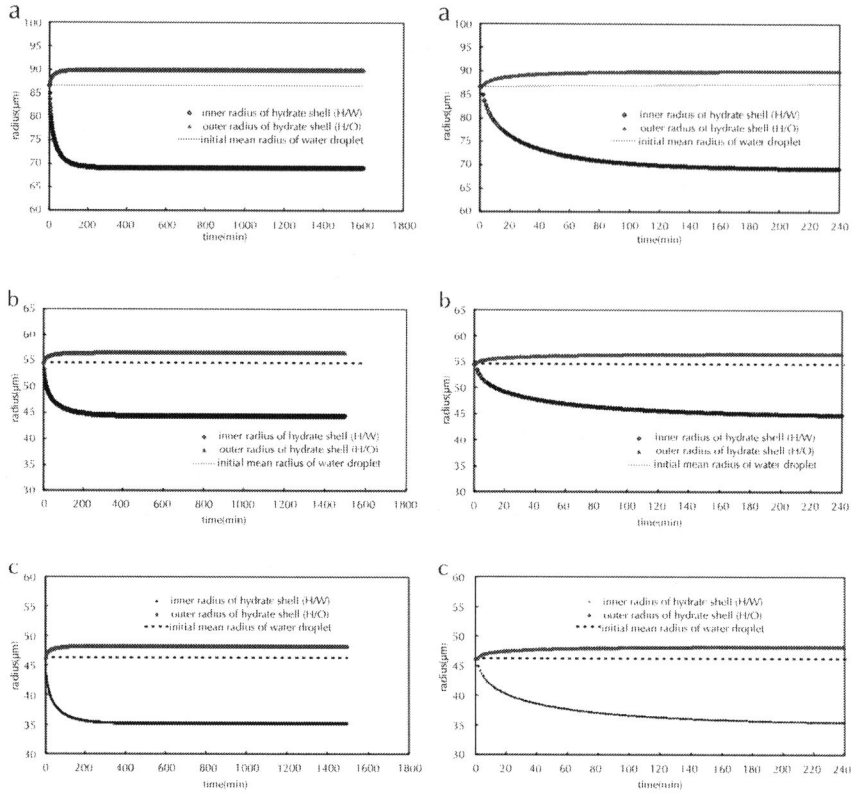

Figure 5: Modeled inner and outer radiuses of hydrate shell as function of time during experiments at (a) 10% water cut, (b) 20% water cut, (c) 25% water cut (2 MPa, 277.15 K).

The variations of the temperatures at H/W and H/O interfaces during the hydrates formation were enormously small according to the calculation by the model at constant temperature boundary condition. It indicated that such heat removal from the hydrate shell balanced with the heat released by hydrates formation.

CONCLUSIONS

The natural gas hydrates formation and growth at 2 MPa and 277.15 K were studied in the flow loop unit for water-in-condensate oil emulsions at different water cuts. The variation of gas consumption as function of time at different water cuts were obtained, and the experimental results showed that the gas consumption value increased with the increment of water cut. The total water conversion rates not only depended on the water cuts, but also related to several other factors, two of which were the amount of the dissolved gas and the surface/volume ratio of the particles. No more gas transformed into hydrates after 3 h, which was likely caused by intrinsic kinetics, mass transfer and heat transfer limitations.

A shell model was proposed for natural gas hydrates growth in condensate oil, in which all the limitations were considered, followed by both inward and outward hydrates growth around the water droplet. The simulated results were in good agreement with the experimental data. The initial radius of the water droplet was a vital parameter, which was defined by the fluid properties and the flow conditions. The dependence on operational conditions was accounted for by adjusting the kinetic constant rate, the initial diffusivity of gas, the initial porous property parameter, and the mass transfer efficiency parameters. The next step towards predictive modeling would correlate these parameters with experimental conditions. For engineering purposes, this model could be very useful to assess the hydrates formation information in the flow loop.

ACKNOWLEDGEMENTS

The authors would like to acknowledge the financial support provide by the National Science & Technology Major Project (no. 2008ZX05026-004-03 and no. 2011ZX05026-004-03) and to thank the anonymous reviewers for their valuable comments and suggestions to improve the quality of the paper.

REFERENCES

1.   E.G. Hammerschmidt, Formation of gas hydrates in natural gas transmission lines, Ind. Eng. Chem. 26 (1934) 851–855.

2.   E.D. Sloan, C.A. Koh, Clathrate Hydrates of Natural Gases, Third ed., CRC Press, Boca Raton, London, New York, 2007.

3.   J.S. Gudmundsson, Cold flow hydrate technology, in: Proc. 4th International Conference on Gas Hydrate, Yokohama, Japan, 2002, pp. 912–916.

4.   D. Turner, L. Talley, Hydrate inhibition via cold flow-no chemicals or insulation, in: Proc. 6th International Conference on Gas Hydrate, Vancouver, British Columbia, Canada, 6–10 July, 2008.

5.   G.J. Chen, T.M. Guo, A new approach to gas hydrate modeling, Chem. Eng. J. 71 (1998) 145–151.

6.   A.L. Ballard, E.D. Sloan, The next generation of hydrate prediction III: Gibbs energy minimization formation, Fluid Phase Equilib. 218 (2004) 15–31.

7.   P. Englezos, N. Kalogerakis, P.D. Dholabhai, P.R. Bishnoi, Kinetics of formation of methane and ethane gas hydrates, Chem. Eng. Sci. 42 (1987) 2647–2658.

8.   P. Englezos, N. Kalogerakis, P.D. Dholabhai, P.R. Bishnoi, Kinetics of hydrate formation form mixtures of methane and ethane, Chem. Eng. Sci. 42 (1987) 2659–2666.

9.   P. Skovborg, P. Rasmussen, A mass transport limited model for the growth of methane and ethane gas hydrates, Chem. Eng. Sci. 49 (1993) 1131–1143.

10.   C.Y. Sun, G.J. Chen, C.F. Ma, Q. Huang, H. Luo, Q.P. Li, The growth kinetics of hydrate film on the surface of gas bubble suspended in water or aqueous surfactant solution, J. Cryst. Growth 306 (2007) 491–499.

11.   Y.H. Mori, T. Mochizuki, Mass transport across clathrate hydrate films-a capillary permeation model, Chem. Eng. Sci. 52 (1997) 3613–3616.

12.   T. Uchida, E.E. Takao, K. Jun'ichi, N. Hideo, Microscopic observation of formation processes of clathrate-hydrate films at an interface between water and carbon dioxide, J. Cryst. Growth 204 (1999) 348–356.

13.   T. Uchida, Y.I. Ikuko, T. Satoshi, E. Takao, N. Jiro, N. Hideo, CO_2 hydrate film formation at the boundary between CO_2 and water: effects of temperature, pressure and additives on the formation rate, J. Cryst. Growth 237 (2002) 383–387.

14.   E.M. Freer, M.S. Selim, E.D. Sloan, Methane hydrate film growth kinetics, Fluid Phase Equilib. 185 (2001) 65–75.

15. Y.H. Mori, Estimating the thickness of hydrate films from their lateral growth rates: application of a simplified heat transfer model, J. Cryst. Growth 223 (2001) 206–212.

16. T. Mochizuki, Y.H. Mori, Clathrate-hydrate film growth along water/hydrateformer phase boundaries-numerical heat-transfer study, J. Cryst. Growth 290 (2006) 642–652.

17. B.Z. Peng, A. Dandekar, C.Y. Sun, H. Luo, Q.L. Ma, W.X. Pang, G.J. Chen, Hydrate film grow on the surface of a gas bubble suspended in water, J. Phys. Chem. B 111 (2007) 12485–12493.

18. T. Mochizuki, Y.H. Mori, Clathrate-hydrate film growth along water/hydrateformer phase boundaries: numerical analyses of mass and heattransfer to/from a hydrate film in relation to its growth, in: Proc. 6th International Conference on Gas Hydrate, Vancouver, British Columbia, Canada, 6–10 July, 2008.

19. D.J. Turner, K.T. Miller, E.D. Sloan, Methane hydrate formation and an inward growing shell model in water-in-oil dispersions, Chem. Eng. Sci. 64 (2009) 3996–4004.

20. D.J. Turner, K.T. Miller, E.D. Sloan, Direct conversionof water droplettomethane hydrate in crude oil, Chem. Eng. Sci. 64 (2009) 5066–5072.

21. A.K. Jamaluddin, N. Kalogerakis, P.R. Bishnoi, Hydrate plugging problems in undersea natural gas pipelines under shutdown conditions, J. Pet. Sci. Eng. 5 (1991) 323–335.

22. D.J. Turner, D.M. Kleehammer, K.T. Miller, C.A. Koh, E.D. Sloan, Formation of hydrate obstructions in pipelines: hydrate particle development and slurry flow, in: Proc. 5th International Conference on Gas Hydrate, Volume 4, Tapir Academic Press, Trondheim, 2005, pp. 1116–1125.

23. J. Gong, B.H. Shi, J.K. Zhao, Natural gas hydrate shell model in gas-slurry pipeline flow, J. Nat. Gas Chem. 19 (2010) 261–266.

24. C.Y. Sun, G.J. Chen, T.M. Guo, R12 hydrate formation kinetics based on laser light scattering technique, Sci. China. Ser. B 46 (2003) 487– 494.

25. C.Y. Sun, G.J. Chen, W.Q. Wang, X.L. Wang, The experimental evaluation for controlling hydrate plug in oil-gas-water multiphase pipeline, October 27–31 Beijing, Proc. 2nd Chinese National Chemical and Biochemical Engineering Annual Meeting (2005).

26.   D.Y. Peng, D.B. Robinson, A new two-constant equation of state, Ind. Eng. Chem. Fundament. 15 (1976) 59–64.

27.   A. Sinquin, T. Palermo, Y. Peysson, Rheological and flow properties of gas hydrate suspensions, Oil Gas Sci. Technol. - Rev. IFP 59 (2004) 41–57.

28.   Y.F. Makogon, Hydrates of Hydrocarbons, PennWell Books, Tulsa, Oklaoma, 1997.

29.   G.J. Chen, C.Y. Sun, Q.L. Ma, Science and Technology of Gas hydrates, first ed., Chemical Industry Press, Beijing, 2007, p. 67.

30.   V. Pauchard, M. Darbouret, T. Palerma, J.-L. Peytavy, Gas hydrate slurry flow in a black oil Prediction of gas hydrate particles agglomeration and linear pressure drop, in: Proc. 13th International Conference on Multiphase Production Technology, Edinburgh, UK, 23–15 June, 2007.

31.   C.Y. Sun, The kinetics ofhydrate formation/ dissociationandrelatedtopics, Ph.D. thesis, China University of Petroleum, Beijing, 2001.

32.   J.K. Zhao, Study on Flow Properties of Hydrate Slurry in Multiphase Pipeline, Ph.D. thesis, China University of Petroleum, Beijing, 2009.

33.   J.D. Lee, R. Susilo, P. Englezos, Methane-ethane and methane-propane hydrate formation and decomposition on water droplets, Chem. Eng. Sci. 60 (2005) 4203–4212.

34.   I.L. Moudrakovski, G.E. McLaurin, C.I. Ratcliffe, J.A. Ripmeester, Methane and carbon dioxide hydrate formation in water droplets: spatially resolved measurements from magnetic resonance microimaging, J. Phys. Chem. B 108 (2004) 17591–17595.

35.   L.W. Zhang, G.J. Chen, C.Y. Sun, S.S. Fan, Y.M. Ding, X.L. Wang, L.Y. Yang, The partition coefficients of ethylene between hydrate and vapor for methane + ethylene + water and methane + ethylene + SDS + water systems, Chem. Eng. Sci. 60 (2005) 5356–5362.

36.   A. Hirata, Y.H. Mori, How liquids wet clathrate hydrates: some macroscopic observations, Chem. Eng. Sci. 53 (1998) 2641–2643.

37.   P.D. Yapa, L. Zheng, F.H. Chen, A model for deepwater oil/gas blowouts, Mar. Pollut. Bull. 43 (2001) 234–241.

38.   Y.F. Makogon, Hydrates of Natural Gas, PennWell Books, Tulsa, Oklaoma, 1981.

Citations

CHAPTER 1

Hongtao Zheng, Yajun Li, and Lin Cai, "Research on Performance of H2 Rich Blowout Limit in Bluff-Body Burner," Mathematical Problems in Engineering, vol. 2012, Article ID 298685, 28 pages, 2012. doi:10.1155/2012/298685.

CHAPTER 2

Yuan Zhu, Guo-ming Chen, Simulation and assessment of SO2 toxic environment after ignition of uncontrolled sour gas flow of well blowout in hills, Journal of Hazardous Materials, Volume 178, Issues 1–3, 15 June 2010, Pages 144-151, ISSN 0304-3894, http://dx.doi.org/10.1016/j.jhazmat.2010.01.055.

CHAPTER 3

P.I. Dolez, C. Nohile, T. Ha Anh, T. Vu-Khanh, R. Benoît, O. Bellavigna-Ladoux, Exploring the chemical aspects of truck tire blowouts and explosions, Safety Science, Volume 46, Issue 9, November 2008, Pages 1334-1344, ISSN 0925-7535, http://dx.doi.org/10.1016/j.ssci.2007.10.004.

CHAPTER 4

Daniel E. Rosner, Manuel Arias-Zugasti, and Michael Labowsky, Intensity and Efficiency of Spray Fuel-Fed Well-Mixed Adiabatic Combustors, doi:10.1016/j.ces.2008.04.027.

CHAPTER 5

Hamed B. Ganji, Reza Ebrahimi, Numerical estimation of blowout, flashback, and flame position in MIT micro gas-turbine chamber, Chemical Engineering Science, Volume 104, 18 December 2013, Pages 857-867, ISSN 0009-2509, http://dx.doi.org/10.1016/j.ces.2013.09.056.

CHAPTER 6

M.R. Talaghat, F. Esmaeilzadeh, D. Mowla, Sand production control by chemical consolidation, Journal of Petroleum Science and Engineering, Volume 67, Issues 1–2, July 2009, Pages 34-40, ISSN 0920-4105, http://dx.doi.org/10.1016/j.petrol.2009.02.005.

CHAPTER 7

L. Popoola, G. Babagana and A. Susu, "A Review of an Expert System Design for Crude Oil Distillation Column Using the Neural Networks Model and Process Optimization and Control Using Genetic Algorithm Framework,"Advances in Chemical Engineering and Science, Vol. 3 No. 2, 2013, pp. 164-170. doi:10.4236/aces.2013.32020.

CHAPTER 8

Bae, J. (2014) Control of Microdomain Orientation in Block Copolymer Thin Films by Electric Field for Proton Exchange Membrane. Advances in Chemical Engineering and Science, 4, 95-102. doi: 10.4236/aces.2014.42013.

CHAPTER 9

Robert B. Barat, Correlating laminar burning velocities using perfectly stirred reactor theory, Chemical Engineering Science, Volume 56, Issue 8, April 2001, Pages 2761-2766, ISSN 0009-2509, http://dx.doi.org/10.1016/S0009-2509(00)00534-0.

CHAPTER 10

Bo-Hui Shi, Jing Gong, Chang-Yu Sun, Jian-Kui Zhao, Yao Ding, and Guang-Jin Chen, An Inward and Outward Natural Gas Hydrates Growth Shell Model Considering Intrinsic Kinetics, Mass and Heat Transfer, doi:10.1016/j.cej.2011.05.029.

Index